Common Core 3 Math Comprehensive Exercise Book

Abundant Math Skill Building Exercises

By

Michael Smith & Reza Nazari

Common Core Exercise Book – Grade 3

Common Core 3 Math Comprehensive Exercise Book

Published in the United State of America By

The Math Notion

Web: WWW.MathNotion.Com

Email: info@Mathnotion.com

Copyright © 2019 by the Math Notion. All rights reserved. No part of this publication may be reproduced, stored in a retrieval system, or transmitted in any form or by any means, electronic, mechanical, photocopying, recording, scanning, or otherwise, except as permitted under Section 107 or 108 of the 1976 United States Copyright Ac, without permission of the author.

All inquiries should be addressed to the Math Notion.

About the Author

Michael Smith has been a math instructor for over a decade now. He holds a master's degree in Management. Since 2006, Michael has devoted his time to both teaching and developing exceptional math learning materials. As a Math instructor and test prep expert, Michael has worked with thousands of students. He has used the feedback of his students to develop a unique study program that can be used by students to drastically improve their math score fast and effectively.

- **SAT Math Comprehensive Exercise Book**
- **ACT Math Comprehensive Exercise Book**
- **GRE Math Comprehensive Exercise Book**
- **SBAC Math Comprehensive Exercise Book**
- **–many Math Education Workbooks, Exercise Books and Study Guides**

As an experienced Math teacher, Mr. Smith employs a variety of formats to help students achieve their goals: He tutors online and in person, he teaches students in large groups, and he provides training materials and textbooks through his website and through Amazon.

You can contact Michael via email at:
info@Mathnotion.com

Common Core Exercise Book – Grade 3

Get ready for the Common Core Math Test with a PERFECT Math Exercise Book!

Common Core Math Comprehensive Exercise Book is designed to help you review all Math topics being covered on the Common Core test and challenge you for achieving high score on your real Common Core Math test. Not only does it provide abundant math exercises, it also contains practice test questions as well as detailed explanations of each answer.

This wide-ranging and updated exercise book covers all Math topics you will ever need to prepare for the Common Core test. It is filled with abundant **math skill building exercises** and worksheets covering fundamental math, arithmetic, pre-algebra, algebra, geometry, basic statistics, probability, and many more math topics. Answers are provided for all questions.

Two full-length Common Core Math tests with detailed explanations can help you improve your knowledge of Mathematics and prepare for the Common Core Math test. This comprehensive exercise book contains many exciting features, including:

- Content 100% aligned with the last updated Common Core test
- 3,000+ Common Core Math practice questions with answers
- Fun and interactive exercises that build confidence
- Designed and developed by Common Core Math test experts
- 2 full-length practice tests (featuring new question types) with detailed answers

After completing this hands-on exercise book, you will gain confidence, strong foundation, and adequate practice to succeed on the Common Core Math test.

Do NOT take the Common Core test without reviewing the Math questions in this Exercise book!

WWW.MathNotion.COM

... So Much More Online!

✓ FREE Math Lessons

✓ More Math Learning Books!

✓ Mathematics Worksheets

✓ Online Math Tutors

For a PDF Version of This Book

Please Visit www.MathNotion.com

Contents

Chapter 1: Place Values and Number Sense 11
    Place Values 12
    Comparing and Ordering Numbers 13
    Write Numbers in Words 14
    Odd or Even 15
    Rounding Numbers 16
    Answers of Worksheets – Chapter 1 17

Chapter 2: Adding and Subtracting 19
    Adding Two-Digit Numbers 20
    Subtracting Two-Digit Numbers 21
    Adding Three-Digit Numbers 22
    Adding Hundreds 23
    Adding 4-Digit Numbers 24
    Subtracting 4-Digit Numbers 25
    Estimate Sums 26
    Estimate Differences 27
    Answers of Worksheets – Chapter 2 28

Chapter 3: Multiplication and Division 31
    Multiplication by 0 to 3 32
    Multiplication by 4 to 7 33
    Multiplication by 8 to 12 34
    Division by 0 to 6 35
    Division by 7 to 12 36
    Dividing by Tens 37
    Divide and Multiply 3-Digit Numbers 38
    Times Table 39
    Answers of Worksheets – Chapter 3 41

Chapter 4: Fractions 43
Fractions of a Number 44
Order Fractions 45
Simplifying Fractions 46
Improper Fractions 47
Comparing Fractions 48
Add Fractions 49
Subtract Fractions 50
Add and Subtract Fractions 51
Compare Sums and Differences of Fractions 52
Add 3 or More Fractions 53
Answers of Worksheets – Chapter 4 54

Chapter 5: Time and Money 59
Read Clocks 60
Digital Clock 61
Measurement – Time 62
Add Money Amounts 63
Subtract Money Amounts 64
Money: Word Problems 65
Answers of Worksheets – Chapter 5 66

Chapter 6: Measurement 67
Metric Length Measurement 68
Customary Length Measurement 68
Metric Capacity Measurement 69
Customary Capacity Measurement 69
Metric Weight and Mass Measurement 70
Customary Weight and Mass Measurement 70
Answers of Worksheets – Chapter 6 71

Chapter 7: Symmetry 73
Line Segments 74
Parallel, Perpendicular and Intersecting Lines 75

Common Core Exercise Book – Grade 3

Identify Lines of Symmetry 76
Lines of Symmetry 77
Answers of Worksheets – Chapter 7 78

Chapter 8: Geometric 81
Identifying Angles 82
Polygon Names 83
Triangles 84
Quadrilaterals and Rectangles 85
Perimeter and Area of Squares 86
Perimeter and Area of rectangles 87
Word Problem 88
Answers of Worksheets – Chapter 8 89

Chapter 9: Patterns and Graphs 91
Repeating Pattern 92
Growing Patterns 93
Patterns: Numbers 94
Bar Graph 95
Tally and Pictographs 96
Line Graphs 97
Answers of Worksheets – Chapter 9 98

Common Core Math Practice Tests 101
Common Core Practice Test 1 105
Common Core Practice Test 2 117

Answers and Explanations 131
Answer Key 131
Practice Test 1 133
Practice Test 2 137

Chapter 1:
Place Values and Number Sense

Topics that you'll practice in this chapter:

✓ Place Values

✓ Compare Numbers

✓ Numbers in Numbers

✓ Rounding

✓ Odd or Even

Common Core Exercise Book – Grade 3

Place Values

✍ Write numbers in expanded form.

1) Thirty–two   30 + 2

2) Sixty–four   ___ + ___

3) Fifty–two   ___ + ___

4) Eighty–six   ___ + ___

5) Ninety–three   ___ + ___

6) Twenty–four   ___ + ___

7) Seventy –four   ___ + ___

8) Forty –one   ___ + ___

9) Ninety–six   ___ + ___

10) Eighty –seven   ___ + ___

11) Thirty –eight   ___ + ___

✍ Circle the correct choice.

12)   The 4 in 84 is in the   ones place   tens place   hundreds place

13)   The 6 in 69 is in the   ones place   tens place   hundreds place

14)   The 2 in 652 is in the   ones place   tens place   hundreds place

15)   The 4 in 540 is in the   ones place   tens place   hundreds place

16)   The 6 in 692 is in the   ones place   tens place   hundreds place

Comparing and Ordering Numbers

Use less than, equal to or greater than.

1) 25 _____ 44
2) 87 _____ 99
3) 47 _____ 35
4) 35 _____ 33
5) 56 _____ 56
6) 58 _____ 52
7) 89 _____ 75
8) 54 _____ 42
9) 23 _____ 23
10) 93 _____ 96
11) 36 _____ 49
12) 69 _____ 58
13) 89 _____ 68
14) 25 _____ 35

Order each set of integers from least to greatest.

15) 9, −11, −5, −3, 7   ___, ___, ___, ___, ___, ___
16) −5, −12, 6, 13, 8   ___, ___, ___, ___, ___, ___
17) 17, −11, −15, 20, −22   ___, ___, ___, ___, ___, ___
18) −15, −25, 22, −8, 42   ___, ___, ___, ___, ___, ___
19) 37, −43, 29, −12, 35   ___, ___, ___, ___, ___, ___
20) 98, 66, −29, 67, −44   ___, ___, ___, ___, ___, ___

Order each set of integers from greatest to least.

21) 10, 14, −2, −14, −8   ___, ___, ___, ___, ___, ___
22) 26, 36, −19, −30, 49   ___, ___, ___, ___, ___, ___
23) 55, −41, −28, 57, −10   ___, ___, ___, ___, ___, ___
24) 78, 91, −24, −20, 84   ___, ___, ___, ___, ___, ___
25) −5, 86, −16, −63, 54   ___, ___, ___, ___, ___, ___
26) −76, −45, −50, 38, 59   ___, ___, ___, ___, ___, ___

Write Numbers in Words

✎ **Write each number in words.**

1) 197 _____

2) 411 _____

3) 556 _____

4) 466 _____

5) 803 _____

6) 718 _____

7) 172 _____

8) 234 _____

9) 902 _____

10) 1,430 _____

11) 1,147 _____

12) 3,374 _____

13) 2,486 _____

14) 1,671 _____

15) 5,260 _____

16) 3,365 _____

17) 3,010 _____

Odd or Even

✎ **Identify whether each number is even or odd.**

1) 16 _____

2) 5 _____

3) 11 _____

4) 24 _____

5) 55 _____

6) 19 _____

7) 36 _____

8) 67 _____

9) 96 _____

10) 18 _____

11) 66 _____

12) 69 _____

✎ **Circle the even number in each group.**

13) 24, 33, 77, 13, 19, 87

14) 11, 19, 87, 53, 33, 48

15) 15, 67, 16, 57, 65, 39

16) 97, 96, 83, 63, 23, 67

✎ **Circle the odd number in each group.**

17) 16, 18, 44, 64, 57, 98

18) 18, 26, 28, 22, 32, 87

19) 48, 84, 97, 94, 62, 58

20) 23, 16, 52, 32, 36, 78

Rounding Numbers

✎ Round each number to the nearest ten.

1) 23   5) 13   9) 48

2) 99   6) 37   10) 62

3) 42   7) 83   11) 78

4) 28   8) 80   12) 65

✎ Round each number to the nearest hundred.

13) 186   17) 223   21) 672

14) 253   18) 312   22) 564

15) 728   19) 288   23) 893

16) 108   20) 928   24) 478

✎ Round each number to the nearest thousand.

25) 1,352   29) 8,039   33) 61,780

26) 2,850   30) 62,536   34) 81,390

27) 5,335   31) 23,432   35) 61,840

28) 4,568   32) 12,845   36) 48,578

Common Core Exercise Book – Grade 3

Answers of Worksheets – Chapter 1

Place Values

1) 30+2
2) 60+4
3) 50+2
4) 80+6
5) 90+3
6) 20+4
7) 70+4
8) 90+1
9) 90+6
10) 80+7
11) 30+8
12) ones place
13) tens place
14) ones place
15) tens place
16) hundreds place

Comparing and Ordering Numbers

1) 25 less than 44
2) 87 less than 99
3) 45 greater than 35
4) 35 greater than 33
5) 56 equals to 56
6) 58 greater than 52
7) 89 greater than 75
8) 54 greater than 42
9) 23 equals to 23
10) 93 less than 96
11) 36 less than 49
12) 69 greater than 58
13) 89 greater than 68
14) 25 less than 35
15) $-11, -5, -3, 7, 9$
16) $-12, -5, 6, 8, 13$
17) $-22, -15, -11, 17, 20$
18) $-25, -15, -8, 22, 42$
19) $-43, -12, 29, 35, 37$
20) $-44, -29, 66, 67, 98$
21) $14, 10, -2, -8, -14$
22) $49, 36, 26, -19, -30$
23) $57, 55, -10, -28, -41$
24) $91, 84, 78, -20, -24$
25) $86, 54, -5, -16, -63$
26) $59, 38, -45, -50, -76$

Write Numbers in Words

1) One hundred ninety-seven
2) Four hundred eleven
3) Five hundred fifty-six
4) Four hundred sixty-six
5) Eight hundred three
6) Seven hundred eighteen
7) one hundred seventy-two
8) Two hundred thirty-four
9) Nine hundred two
10) One thousand, four hundred thirty
11) One thousand, 0ne hundred forty-seven

12) Three thousand, three hundred seventy-four
13) Two thousand, four hundred eighty-six
14) one thousand, six hundred seventy-one
15) Five thousand, two hundred sixty
16) Three thousand, three hundred sixty-five
17) Three thousand, ten

Odd or Even

1) Even	6) Odd	11) Even	16) 96
2) Odd	7) Even	12) Odd	17) 57
3) Odd	8) Odd	13) 24	18) 87
4) Even	9) Even	14) 48	19) 97
5) Odd	10) Even	15) 16	20) 23

Rounding Numbers

1) 20	13) 200		25) 1,000
2) 100	14) 300		26) 3,000
3) 40	15) 700		27) 5,000
4) 30	16) 100		28) 5,000
5) 10	17) 200		29) 8,000
6) 40	18) 300		30) 62,000
7) 80	19) 300		31) 23,000
8) 80	20) 900		32) 13,000
9) 50	21) 700		33) 62,000
10) 60	22) 600		34) 81,000
11) 80	23) 900		35) 62,000
12) 70	24) 500		36) 49,000

Chapter 2: Adding and Subtracting

Topics that you'll practice in this chapter:

- ✓ Adding Two–Digit Numbers

- ✓ Subtracting Two–Digit Numbers

- ✓ Adding Three–Digit Numbers

- ✓ Adding Hundreds

- ✓ Adding 4–Digit Numbers

- ✓ Subtracting 4–Digit Numbers

- ✓ Estimate Sums

- ✓ Estimate Differences

Adding Two-Digit Numbers

✎ **Find each sum.**

1) 40
  + 28
  ———

2) 42
  + 24
  ———

3) 36
  + 11
  ———

4) 22
  + 22
  ———

5) 55
  + 20
  ———

6) 34
  + 25
  ———

7) 79
  + 6
  ———

8) 43
  + 12
  ———

9) 80
  + 20
  ———

10) 24
   + 12
   ———

11) 40
   + 22
   ———

12) 29
   + 18
   ———

13) 19
   + 35
   ———

14) 39
   + 21
   ———

15) 55
   + 33
   ———

16) 63
   + 30
   ———

17) 66
   + 36
   ———

18) 38
   + 23
   ———

Subtracting Two-Digit Numbers

✍ **Find each difference.**

1) 16 − 10

2) 60 − 16

3) 87 − 27

4) 28 − 11

5) 53 − 12

6) 69 − 20

7) 79 − 32

8) 86 − 18

9) 67 − 21

10) 68 − 17

11) 69 − 26

12) 36 − 11

13) 68 − 29

14) 59 − 36

15) 39 − 11

16) 93 − 21

17) 59 − 35

18) 96 − 63

Adding Three–Digit Numbers

✎ **Find each sum.**

1) 426
   + 36
   -----

2) 625
   + 130
   -----

3) 325
   + 153
   -----

4) 563
   + 125
   -----

5) 453
   + 230
   -----

6) 298
   + 120
   -----

7) 689
   + 56
   -----

8) 863
   + 325
   -----

9) 865
   + 65
   -----

10) 269
   + 120
   -----

11) 187
   + 125
   -----

12) 289
   + 150
   -----

13) 369
   + 156
   -----

14) 360
   + 150
   -----

15) 589
   + 263
   -----

16) 890
   + 345
   -----

17) 620
   + 215
   -----

18) 680
   + 230

Adding Hundreds

Add.

1) 200 + 100 = - - -

2) 100 + 500 = - - -

3) 300 + 300 = - - -

4) 300 + 600 = - - -

5) 100 + 100 = - - -

6) 400 + 200 = - - -

7) 300 + 700 = - - -

8) 600 + 300 = - - -

9) 200 + 700 = - - -

10) 500 + 800 = - - -

11) 200 + 800 = - - -

12) 600 + 400 = - - -

13) 600 + 900 = - - -

14) 100 + 800 = - - -

15) 900 + 100 = - - -

16) 300 + 900 = - - -

17) 700 + 100 = - - -

18) 200 + 500 = - - -

19) 100 + 600 = - - -

20) 400 + 400 = - - -

21) 600 + 600 = - - -

22) 800 + 100 = - - -

23) 800 + 700 = - - -

24) 900 + 800 = - - -

25) If there are 800 balls in a box and Jackson puts 400 more balls inside, how many balls are in the box? _____ balls

Common Core Exercise Book – Grade 3

Adding 4–Digit Numbers

Add.

27) 1,135
  + 5,236
  _____

30) 2,125
  +4,035
  _____

33) 3,236
  +2,369
  _____

28) 3,369
  + 1,356
  _____

31) 3,135
  +2,194
  _____

34) 6,320
  +3,765
  _____

29) 5,598
  + 2,325
  _____

32) 4,036
  +2,365
  _____

35) 3,890
  +3,567
  _____

Find the missing numbers.

10) 1,145 + ___ = 1,369

13) 655 + ___ = 1,986

11) 600 + 2,000 = ___

14) ___ + 820 = 1,450

12) 4,200 + ___ = 6,300

15) ___ + 1670 = 3,230

16) Mason sells gems. He finds a diamond in Istanbul and buys it for $4,433. Then, he flies to Cairo and purchases a bigger diamond for the bargain price of $8,922. How much does Mason spend on the two diamonds?

Subtracting 4-Digit Numbers

✎ **Subtract.**

1) 3,130 − 1,135

2) 4,356 − 1,870

3) 6,986 − 3,678

4) 5,987 − 5,422

5) 6,362 − 4,331

6) 8,365 − 3,212

7) 9,356 − 6,712

8) 9,350 − 3,729

9) 7,117 − 2,216

✎ **Find the missing number.**

10) 3,223 − __ = 1,320

11) 4,856 − __ = 3,245

12) 1,136 − 689 = __

13) 3,200 − __ = 1,450

14) 4,870 − 1,650 = __

15) 5,360 − 3,320 = __

16) Bob had $4,486 invested in the stock market until he lost $3,198 on those investments. How much money does he have in the stock market now?

Estimate Sums

✎ **Estimate the sum by rounding each added to the nearest ten.**

1) 35 + 8 =

2) 27 + 45 =

3) 35 + 13 =

4) 36 + 39 =

5) 12 + 35 =

6) 36 + 12 =

7) 48 + 25 =

8) 35 + 78 =

9) 45 + 86 =

10) 63 + 57 =

11) 45 + 36 =

12) 51 + 15 =

13) 35 + 58 =

14) 36 + 65 =

15) 86 + 84 =

16) 16 + 69 =

17) 65 + 64 =

18) 31 + 26 =

19) 71 + 48 =

20) 35 + 64 =

21) 12 + 93 =

22) 62 + 52 =

23) 163 + 142 =

24) 53 + 75 =

Estimate Differences

✎ **Estimate the difference by rounding each number to the nearest ten.**

1) 56 − 22 =

2) 33 − 24 =

3) 78 − 46 =

4) 42 − 23 =

5) 69 − 46 =

6) 44 − 22 =

7) 77 − 47 =

8) 48 − 29 =

9) 94 − 48 =

10) 78 − 58 =

11) 68 − 26 =

12) 82 − 37 =

13) 73 − 43 =

14) 59 − 42 =

15) 84 − 53 =

16) 65 − 42 =

17) 97 − 84 =

18) 42 − 22 =

19) 56 − 49 =

20) 89 − 26 =

21) 84 − 68 =

22) 65 − 13 =

23) 77 − 7 =

24) 76 − 33 =

Answers of Worksheets – Chapter 2

Adding two–digit numbers

1) 68
2) 66
3) 47
4) 44
5) 75
6) 59
7) 85
8) 55
9) 100
10) 36
11) 62
12) 47
13) 54
14) 60
15) 88
16) 93
17) 102
18) 61

Subtracting two–digit numbers

1) 6
2) 44
3) 60
4) 17
5) 41
6) 49
7) 47
8) 68
9) 46
10) 51
11) 43
12) 25
13) 39
14) 23
15) 28
16) 72
17) 24
18) 33

Adding three–digit numbers

1) 462
2) 755
3) 478
4) 688
5) 683
6) 418
7) 745
8) 1,188
9) 930
10) 389
11) 312
12) 439
13) 525
14) 510
15) 852
16) 1,235
17) 835
18) 910

Adding hundreds

1) 300
2) 600
3) 600
4) 900
5) 200
6) 600
7) 1,000
8) 900
9) 900
10) 1,300
11) 1,000
12) 1,000
13) 1,500
14) 900
15) 1,000
16) 1,200
17) 800
18) 700
19) 700
20) 800
21) 1,200

22) 900
23) 1,500
24) 1,700
25) 1,200

Adding 4–digit numbers

1) 6,371
2) 4,725
3) 7,923
4) 6,160
5) 5,329
6) 6401
7) 5,605
8) 10,085
9) 7,457
10) 224
11) 2,600
12) 2,100
13) 1,331
14) 630
15) 1,560
16) $13,355

Subtracting 4–digit numbers

1) 1,995
2) 2,486
3) 3,308
4) 565
5) 2,031
6) 5,153
7) 2,644
8) 5,621
9) 4,901
10) 1,903
11) 1,611
12) 447
13) 1,750
14) 3,220
15) 2,040
16) 1,288

Estimate sums

1) 50
2) 80
3) 50
4) 80
5) 50
6) 50
7) 80
8) 120
9) 140
10) 120
11) 90
12) 70
13) 100
14) 110
15) 170
16) 90
17) 130
18) 60
19) 120
20) 100
21) 100
22) 110
23) 300
24) 130

Estimate differences

1) 40
2) 10
3) 30
4) 20
5) 20
6) 20
7) 30
8) 20
9) 40
10) 20
11) 40
12) 40
13) 30
14) 20
15) 30
16) 30
17) 20
18) 20
19) 10
20) 60
21) 10
22) 60
23) 70
24) 50

Chapter 3: Multiplication and Division

Topics that you'll practice in this chapter:

✓ Times Table

✓ Multiplication by 0 to 3

✓ Multiplication by 4 to 7

✓ Multiplication by 8 to 12

✓ Division by 0 to 6

✓ Division by 7 to 12

✓ Dividing by Tens

✓ Divide and Multiply 3–Digit Numbers by 1-digit Numbers

Multiplication by 0 to 3

✎ Write the answers.

1) $4 \times 3 =$ ___

2) $2 \times 3 =$ ___

3) $6 \times 3 =$ ___

4) $3 \times 8 =$ ___

5) $3 \times 6 =$ ___

6) $2 \times 6 =$ ___

7) $10 \times 2 =$ ___

8) $3 \times 11 =$ ___

9) $10 \times 3 =$ ___

10) $2 \times 0 =$ ___

11) $8 \times 2 =$ ___

12) $12 \times 3 =$ ___

13) $9 \times 3 =$ ___

14) $12 \times 2 =$ ___

15) $11 \times 1 =$ ___

16) $9 \times 2 =$ ___

✎ *Find Each Missing Number.*

17) $2 \times$ ___ $= 16$

18) $4 \times$ ___ $= 12$

19) $5 \times 3 =$ ___

20) ___ $\times 2 = 10$

21) ___ $\times 3 = 27$

22) $3 \times$ ___ $= 0$

23) $2 \times$ ___ $= 24$

24) $7 \times$ ___ $= 21$

25) $2 \times$ ___ $= 18$

26) ___ $\times 2 = 14$

27) ___ $\times 3 = 9$

28) $2 \times 0 =$ ___

29) $2 \times$ ___ $= 20$

30) ___ $\times 2 = 22$

Multiplication by 4 to 7

Write the answers.

1) 7 × 4 = ___

2) 5 × 4 = ___

3) 10 × 4 = ___

4) 12 × 5 = ___

5) 9 × 5 = ___

6) 20 × 5 = ___

7) 8 × 7 = ___

8) 6 × 8 = ___

9) 2 × 5 = ___

10) 0 × 5 = ___

11) 12 × 6 = ___

12) 8 × 4 = ___

13) 8 × 2 = ___

14) 13 × 5 = ___

15) 7 × 3 = ___

16) 8 × 1 = ___

17) 10 × 6 = ___

18) 6 × 4 = ___

19) 11 × 6 = ___

20) 0 × 7 = ___

21) 1 × 6 = ___

22) 10 × 7 = ___

23) 12 × 4 = ___

24) 12 × 7 = ___

25) 11 × 5 = ___

26) 11 × 7 = ___

27) Ryan ordered seven pizzas and sliced them into four pieces each. How many pieces of pizza were there?

_____

Multiplication by 8 to 12

✎ **Write the answers.**

1) 11 × 5 = ___

2) 9 × 6 = ___

3) 4 × 9 = ___

4) 10 × 10 = ___

5) 11 × 7 = ___

6) 8 × 9 = ___

7) 12 × 12 = ___

8) 9 × 11 = ___

9) 9 × 7 = ___

10) 10 × 7 = ___

11) 8 × 11 = ___

12) 9 × 10 = ___

13) 5 × 10 = ___

14) 4 × 12 = ___

15) 9 × 9 = ___

16) 2 × 9 = ___

17) 11 × 3 = ___

18) 11 × 11 = ___

19) 10 × 12 = ___

20) 12 × 7 = ___

21) 8 × 12 = ___

22) 11 × 10 = ___

23) There are 10 bananas in each box. How many bananas are in 9 boxes?

24) Each child has 9 apples. If there are 6 children, how many apples are there in total?

25) Each child has 11 pencils. If there are 12 children, how many pens are there in total?

Division by 0 to 6

✎ **Find each missing number.**

1) $44 \div \underline{} = 11$

2) $24 \div 6 = \underline{}$

3) $12 \div 4 = \underline{}$

4) $36 \div 6 = \underline{}$

5) $\underline{} \div 3 = 11$

6) $20 \div 2 = \underline{}$

7) $\underline{} \div 3 = 9$

8) $18 \div \underline{} = 6$

9) $\underline{} \div 2 = 16$

10) $\underline{} \div 6 = 12$

11) $\underline{} \div 5 = 1$

12) $18 \div 2 = \underline{}$

13) $8 \div \underline{} = 2$

14) $14 \div 7 = \underline{}$

15) $36 \div \underline{} = 6$

16) $\underline{} \div 3 = 8$

17) $40 \div 5 = \underline{}$

18) $48 \div \underline{} = 12$

19) $25 \div \underline{} = 5$

20) $32 \div \underline{} = 8$

21) $42 \div 6 = \underline{}$

22) $100 \div 5 = \underline{}$

23) $\underline{} \div 4 = 4$

24) $36 \div \underline{} = 12$

25) $50 \div 2 = \underline{}$

26) $3 \div \underline{} = 1$

27) $100 \div \underline{} = 20$

28) $60 \div \underline{} = 15$

29) $24 \div \underline{} = 12$

30) $\underline{} \div 6 = 10$

31) $36 \div 4 = \underline{}$

32) $20 \div 5 = \underline{}$

33) $\underline{} \div 5 = 7$

34) $45 \div 5 = \underline{}$

35) $54 \div 6 = \underline{}$

36) $\underline{} \div 7 = 9$

37) Mia has 50 strawberries that she would like to give to her 5 friends. If she shares them equally, how many strawberries will she give to each of her friends?

Division by 7 to 12

✎ **Find each missing number.**

1) ___ ÷ 12 = 1

2) ___ ÷ 8 = 5

3) 77 ÷ 7 = ___

4) 99 ÷ ___ = 11

5) 96 ÷ 8 = ___

6) ___ ÷ 8 = 8

7) 72 ÷ ___ = 9

8) ___ ÷ 7 = 9

9) 27 ÷ ___ = 3

10) 42 ÷ 7 = ___

11) ___ ÷ 3 = 9

12) 81 ÷ ___ = 9

13) ___ ÷ 7 = 7

14) 80 ÷ 8 = ___

15) 121 ÷ ___ = 11

16) 56 ÷ 7 = ___

17) 110 ÷ 10 = ___

18) 108 ÷ 9 = ___

19) 20 ÷ ___ = 2

20) 72 ÷ 12 = ___

21) 77 ÷ ___ = 7

22) 120 ÷ 10 = ___

23) 132 ÷ ___ = 11

24) 32 ÷ ___ = 4

25) ___ ÷ 10 = 9

26) 110 ÷ 11 = ___

27) 120 ÷ ___ = 10

28) 70 ÷ ___ = 7

29) ___ ÷ 11 = 8

30) ___ ÷ 12 = 12

31) 108 ÷ ___ = 12

32) ___ ÷ 11 = 2

33) 84 ÷ 12 = ___

34) 100 ÷ 10 = ___

35) 7 ÷ 7 = ___

36) 66 ÷ ___ = 6

37) Stella has 49 fruit juice that she would like to give to her 7 friends. If she shares them equally, how many fruit juices will she give to each?

Dividing by Tens

✎ Find answers.

1) $400 \div 10 =$ ___

2) $800 \div 20 =$ ___

3) $1200 \div 40 =$ ___

4) $350 \div 10 =$ ___

5) $420 \div 60 =$ ___

6) $200 \div 40 =$ ___

7) $350 \div 50 =$ ___

8) $640 \div 80 =$ ___

9) $630 \div 70 =$ ___

10) $320 \div 80 =$ ___

11) $720 \div 80 =$ ___

12) $160 \div 40 =$ ___

13) $250 \div 50 =$ ___

14) $490 \div 70 =$ ___

15) $150 \div 30 =$ ___

16) $560 \div 80 =$ ___

17) $240 \div 30 =$ ___

18) $560 \div 20 =$ ___

19) $\frac{520}{20} =$ ___

20) $\frac{600}{20} =$ ___

21) $\frac{900}{90} =$ ___

22) $\frac{630}{30} =$ ___

23) $\frac{270}{30} =$ ___

24) $\frac{320}{40} =$ ___

Divide and Multiply 3–Digit Numbers

Find the answers.

1) 350 ÷ 5 = ____
2) 180 ÷ 6 = ____
3) 270 ÷ 3 = ____
4) 900 ÷ 9 = ____
5) 630 ÷ 9 = ____
6) 810 ÷ 3 = ____
7) 800 ÷ 8 = ____
8) 450 ÷ 5 = ____
9) 210 ÷ 7 = ____
10) 720 ÷ 8 = ____

11) 200 ÷ 4 = ____
12) 480 ÷ 8 = ____
13) 840 ÷ 4 = ____
14) 420 ÷ 7 = ____
15) 100 ÷ 4 = ____
16) 800 ÷ 4 = ____
17) 300 ÷ 6 = ____
18) 480 ÷ 4 = ____
19) 280 ÷ 7 = ____
20) 150 ÷ 6 = ____

Find the answers.

21) 420 × 4

22) 150 × 5

23) 430 × 6

24) 270 × 3

25) 870 × 3

26) 532 × 9

27) 656 × 7

28) 305 × 4

29) 259 × 8

Times Table

×	1	2	3	4	5	6	7	8	9	10	11	12
1	1	2	3	4	5	6	7	8	9	10	11	12
2	2	4	6	8	10	12	14	16	18	20	22	24
3	3	6	9	12	15	18	21	24	27	30	33	36
4	4	8	12	16	20	24	(28)	32	36	40	44	48
5	5	10	15	20	25	30	35	40	45	50	55	60
6	6	12	18	24	30	36	42	48	54	60	66	72
7	7	14	21	28	35	42	49	56	63	70	77	84
8	8	16	24	32	40	48	56	64	72	80	88	96
9	9	18	27	36	45	54	63	72	(81)	90	99	108
10	10	20	30	40	50	60	70	80	90	100	110	120
11	11	22	33	44	55	66	77	88	99	110	121	132
12	12	24	36	48	60	72	84	96	108	120	132	144

Answers of Worksheets – Chapter 3

Multiplication by 0 to 3

1) 12
2) 6
3) 18
4) 24
5) 18
6) 12
7) 20
8) 33
9) 30
10) 0
11) 16
12) 36
13) 27
14) 24
15) 11
16) 18
17) 8
18) 3
19) 15
20) 5
21) 9
22) 0
23) 12
24) 3
25) 9
26) 7
27) 3
28) 0
29) 10
30) 11

Multiplication by 4 to 7

1) 28
2) 20
3) 40
4) 60
5) 45
6) 100
7) 56
8) 48
9) 10
10) 0
11) 72
12) 32
13) 16
14) 65
15) 21
16) 8
17) 60
18) 24
19) 66
20) 0
21) 6
22) 70
23) 48
24) 84
25) 55
26) 77
27) 28

Multiplication by 8 to 12

1) 55
2) 54
3) 36
4) 100
5) 77
6) 72
7) 144
8) 99
9) 63
10) 70
11) 88
12) 90
13) 50
14) 48
15) 81
16) 18
17) 33
18) 121
19) 120
20) 84
21) 96
22) 110
23) 90
24) 54
25) 132

Division by 0 to 6

1) 4
2) 4
3) 3
4) 6
5) 33
6) 10
7) 27
8) 3
9) 32
10) 72
11) 5
12) 9
13) 4
14) 2
15) 6
16) 24
17) 8
18) 4
19) 5
20) 4

21) 7	26) 3	31) 9	36) 63
22) 20	27) 5	32) 4	37) 10
23) 16	28) 4	33) 35	
24) 3	29) 2	34) 9	
25) 25	30) 60	35) 9	

Division by 7 to 12

1) 12	11) 27	21) 11	31) 9
2) 40	12) 9	22) 12	32) 22
3) 11	13) 49	23) 12	33) 7
4) 9	14) 10	24) 8	34) 10
5) 12	15) 11	25) 90	35) 1
6) 64	16) 8	26) 10	36) 11
7) 8	17) 11	27) 12	37) 7
8) 63	18) 12	28) 10	
9) 9	19) 10	29) 88	
10) 6	20) 6	30) 144	

Dividing by tens

1) 40	7) 7	13) 5	19) 26
2) 40	8) 8	14) 7	20) 30
3) 30	9) 9	15) 5	21) 10
4) 35	10) 7	16) 7	22) 21
5) 7	11) 9	17) 8	23) 9
6) 5	12) 4	18) 28	24) 8

Divide and Multiply 3-digit numbers by 1-digit numbers

1) 70	9) 30	17) 50	25) 2,610
2) 30	10) 90	18) 120	26) 4,788
3) 90	11) 50	19) 40	27) 4592
4) 100	12) 60	20) 25	28) 1,220
5) 70	13) 210	21) 1,680	29) 2,072
6) 270	14) 60	22) 750	
7) 100	15) 25	23) 2,580	
8) 90	16) 200	24) 810	

Chapter 4: Fractions

Topics that you'll learn in this chapter:

- ✓ Fractions
- ✓ Fractions of a Number
- ✓ Order Fractions
- ✓ Simplifying Fractions
- ✓ Improper Fractions
- ✓ Comparing Fractions
- ✓ Add Fractions
- ✓ Subtract Fractions
- ✓ Compare Sums and Differences of Fractions
- ✓ Add 3 or More Fractions

Fractions of a Number

Solve.

1) Find $\frac{1}{2}$ of 80.

2) Find $\frac{4}{9}$ of 180.

3) Find $\frac{1}{2}$ of 72.

4) Find $\frac{3}{4}$ of 8.

5) Find $\frac{1}{40}$ of 80.

6) Find $\frac{3}{2}$ of 30.

7) Find $\frac{1}{3}$ of 21.

8) Find $\frac{2}{3}$ of 45.

9) Find $\frac{1}{4}$ of 12.

10) Find $\frac{3}{7}$ of 28.

11) Find $\frac{7}{5}$ of 30.

12) Find $\frac{2}{9}$ of 27.

13) Find $\frac{1}{6}$ of 60.

14) Find $\frac{1}{3}$ of 18.

15) Find $\frac{3}{4}$ of 124.

16) Find $\frac{5}{8}$ of 32.

17) Find $\frac{7}{6}$ of 360.

18) Find $\frac{4}{10}$ of 200.

19) Find $\frac{3}{4}$ of 32.

20) Find $\frac{1}{20}$ of 40.

Order Fractions

✎ Order the fractions from greatest to latest.

1) $\dfrac{1}{4}, \dfrac{4}{3}, \dfrac{1}{8}$

2) $\dfrac{1}{5}, \dfrac{14}{5}, \dfrac{1}{8}$

3) $\dfrac{6}{10}, \dfrac{5}{4}, \dfrac{1}{10}$

4) $\dfrac{3}{4}, \dfrac{3}{2}, \dfrac{2}{7}$

5) $\dfrac{1}{3}, \dfrac{3}{4}, \dfrac{3}{10}$

6) $\dfrac{1}{4}, \dfrac{1}{2}, \dfrac{3}{4}$

7) $\dfrac{1}{2}, \dfrac{1}{10}, \dfrac{7}{10}$

8) $\dfrac{3}{4}, \dfrac{9}{4}, \dfrac{5}{4}$

9) $\dfrac{5}{7}, \dfrac{2}{7}, \dfrac{3}{7}$

10) $\dfrac{8}{5}, \dfrac{1}{8}, \dfrac{5}{8}$

✎ Order the fractions from latest to greatest.

11) $\dfrac{7}{4}, \dfrac{5}{4}, \dfrac{8}{4}$

12) $\dfrac{5}{7}, \dfrac{8}{7}, \dfrac{4}{7}$

13) $\dfrac{5}{6}, \dfrac{3}{7}, \dfrac{1}{6}$

14) $\dfrac{1}{5}, \dfrac{3}{5}, \dfrac{2}{5}$

15) $\dfrac{3}{6}, \dfrac{5}{6}, \dfrac{1}{6}$

16) $\dfrac{8}{3}, \dfrac{5}{3}, \dfrac{3}{4}$

17) $\dfrac{11}{5}, \dfrac{2}{5}, \dfrac{8}{5}$

18) $\dfrac{17}{3}, \dfrac{22}{3}, \dfrac{11}{3}$

Simplifying Fractions

Simplify each fraction to its lowest terms.

1) $\dfrac{18}{36} =$

2) $\dfrac{16}{20} =$

3) $\dfrac{18}{24} =$

4) $\dfrac{15}{60} =$

5) $\dfrac{36}{48} =$

6) $\dfrac{24}{36} =$

7) $\dfrac{24}{30} =$

8) $\dfrac{8}{32} =$

9) $\dfrac{36}{72} =$

10) $\dfrac{12}{18} =$

11) $\dfrac{26}{78} =$

12) $\dfrac{42}{56} =$

13) $\dfrac{126}{154} =$

14) $\dfrac{108}{120} =$

15) $\dfrac{9}{12} =$

16) $\dfrac{20}{84} =$

17) $\dfrac{75}{180} =$

18) $\dfrac{90}{150} =$

19) $\dfrac{64}{136} =$

20) $\dfrac{150}{650} =$

21) $\dfrac{240}{480} =$

Solve each problem.

22) Which of the following fractions equal to $\dfrac{7}{6}$? _____

A. $\dfrac{24}{72}$ B. $\dfrac{84}{72}$ C. $\dfrac{63}{48}$ D. $\dfrac{70}{72}$

23) Which of the following fractions equal to $\dfrac{2}{7}$? _____

A. $\dfrac{38}{133}$ B. $\dfrac{32}{287}$ C. $\dfrac{36}{260}$ D. $\dfrac{39}{84}$

24) Which of the following fractions equal to $\dfrac{2}{9}$? _____

A. $\dfrac{84}{386}$ B. $\dfrac{120}{540}$ C. $\dfrac{40}{108}$ D. $\dfrac{109}{525}$

Improper Fractions

✍ Fill in the blank.

1) $\dfrac{1}{3} + \_\_ = 1$

2) $\dfrac{4}{5} + \_\_ = 5$

3) $\dfrac{1}{5} + \_\_ = 1$

4) $\dfrac{3}{2} + \_\_ = 2$

5) $\dfrac{4}{7} + \_\_ = 1$

6) $\dfrac{1}{4} + \_\_ = 2$

7) $\dfrac{1}{5} + \_\_ = 3$

8) $\dfrac{3}{4} + \_\_ = 4$

9) $\dfrac{3}{7} + \_\_ = 2$

10) $\dfrac{1}{6} + \_\_ = 1$

✍ Convert Improper fractions to mixed numbers.

11) $\dfrac{5}{4} =$

12) $\dfrac{7}{2} =$

13) $\dfrac{7}{5} =$

14) $\dfrac{13}{2} =$

15) $\dfrac{7}{4} =$

16) $\dfrac{13}{3} =$

17) $\dfrac{7}{3} =$

18) $\dfrac{10}{4} =$

19) $\dfrac{12}{8} =$

20) $\dfrac{14}{5} =$

Comparing Fractions

Use > = < to compare fractions.

1) $\dfrac{1}{3} \square \dfrac{3}{9}$

2) $\dfrac{4}{24} \square \dfrac{1}{4}$

3) $\dfrac{24}{48} \square \dfrac{4}{16}$

4) $\dfrac{8}{10} \square \dfrac{4}{5}$

5) $\dfrac{15}{45} \square \dfrac{1}{3}$

6) $\dfrac{21}{28} \square \dfrac{6}{5}$

7) $\dfrac{6}{18} \square \dfrac{12}{36}$

8) $\dfrac{11}{22} \square \dfrac{20}{22}$

Find the missing values.

9) $\dfrac{1}{4} = \dfrac{}{20}$

10) $\dfrac{}{4} = \dfrac{4}{16}$

11) $\dfrac{}{9} = \dfrac{1}{3}$

12) $\dfrac{4}{12} = \dfrac{1}{}$

13) $\dfrac{}{12} = \dfrac{20}{48}$

14) $\dfrac{9}{72} = \dfrac{}{8}$

15) $\dfrac{3}{16} = \dfrac{6}{}$

16) $\dfrac{7}{10} = \dfrac{21}{}$

17) $\dfrac{3}{18} = \dfrac{}{6}$

18) $\dfrac{4}{32} = \dfrac{}{8}$

Add Fractions

✎ **Add fractions.**

1) $\dfrac{3}{4} + \dfrac{1}{4} =$

2) $\dfrac{3}{7} + \dfrac{4}{7} =$

3) $\dfrac{5}{7} + \dfrac{4}{7} =$

4) $\dfrac{5}{2} + \dfrac{5}{2} =$

5) $\dfrac{4}{11} + \dfrac{3}{11} =$

6) $\dfrac{3}{8} + \dfrac{4}{8} =$

7) $\dfrac{7}{5} + \dfrac{4}{5} =$

8) $\dfrac{5}{13} + \dfrac{6}{13} =$

9) $\dfrac{5}{17} + \dfrac{10}{17} =$

10) $\dfrac{3}{9} + \dfrac{5}{9} =$

11) $\dfrac{5}{14} + \dfrac{6}{14} =$

12) $\dfrac{6}{21} + \dfrac{10}{21} =$

13) $\dfrac{8}{13} + \dfrac{6}{13} =$

14) $\dfrac{4}{19} + \dfrac{3}{19} =$

15) $\dfrac{7}{15} + \dfrac{3}{15} =$

16) $\dfrac{15}{35} + \dfrac{12}{35} =$

17) $\dfrac{7}{25} + \dfrac{10}{25} =$

18) $\dfrac{6}{22} + \dfrac{8}{22} =$

19) $\dfrac{20}{43} + \dfrac{11}{43} =$

20) $\dfrac{8}{39} + \dfrac{15}{39} =$

21) $\dfrac{20}{71} + \dfrac{16}{71} =$

22) $\dfrac{32}{52} + \dfrac{10}{52} =$

Subtract Fractions

✎ **Subtract fractions.**

1) $\dfrac{3}{7} - \dfrac{2}{7} =$

2) $\dfrac{2}{5} - \dfrac{1}{5} =$

3) $\dfrac{14}{18} - \dfrac{8}{18} =$

4) $\dfrac{7}{9} - \dfrac{3}{9} =$

5) $\dfrac{4}{12} - \dfrac{3}{12} =$

6) $\dfrac{5}{15} - \dfrac{3}{15} =$

7) $\dfrac{7}{6} - \dfrac{5}{6} =$

8) $\dfrac{11}{15} - \dfrac{7}{15} =$

9) $\dfrac{20}{25} - \dfrac{7}{25} =$

10) $\dfrac{7}{24} - \dfrac{6}{24} =$

11) $\dfrac{15}{31} - \dfrac{12}{31} =$

12) $\dfrac{10}{28} - \dfrac{9}{28} =$

13) $\dfrac{9}{25} - \dfrac{6}{25} =$

14) $\dfrac{35}{40} - \dfrac{17}{40} =$

15) $\dfrac{22}{20} - \dfrac{9}{20} =$

16) $\dfrac{26}{40} - \dfrac{18}{40} =$

17) $\dfrac{32}{23} - \dfrac{28}{23} =$

18) $\dfrac{16}{82} - \dfrac{6}{82} =$

19) $\dfrac{35}{40} - \dfrac{15}{40} =$

20) $\dfrac{29}{35} - \dfrac{19}{35} =$

21) $\dfrac{21}{60} - \dfrac{11}{60} =$

22) $\dfrac{9}{28} - \dfrac{8}{28} =$

Add and Subtract Fractions

✎ Add fractions.

1) $\dfrac{1}{3} + \dfrac{2}{3} =$

2) $\dfrac{1}{5} + \dfrac{4}{5} =$

3) $\dfrac{4}{7} + \dfrac{2}{7} =$

4) $\dfrac{5}{9} + \dfrac{2}{9} =$

5) $\dfrac{3}{12} + \dfrac{8}{12} =$

6) $\dfrac{4}{20} + \dfrac{1}{20} =$

7) $\dfrac{3}{9} + \dfrac{2}{9} =$

8) $\dfrac{4}{6} + \dfrac{2}{6} =$

9) $\dfrac{1}{24} + \dfrac{1}{24} =$

10) $\dfrac{16}{23} + \dfrac{5}{23} =$

✎ Subtract fractions.

11) $\dfrac{4}{7} - \dfrac{2}{7} =$

12) $\dfrac{6}{8} - \dfrac{5}{8} =$

13) $\dfrac{3}{11} - \dfrac{2}{11} =$

14) $\dfrac{10}{18} - \dfrac{3}{18} =$

15) $\dfrac{9}{19} - \dfrac{4}{19} =$

16) $\dfrac{3}{13} - \dfrac{1}{13} =$

17) $\dfrac{15}{33} - \dfrac{13}{33} =$

18) $\dfrac{25}{55} - \dfrac{23}{55} =$

19) $\dfrac{19}{61} - \dfrac{17}{61} =$

20) $\dfrac{12}{82} - \dfrac{7}{82} =$

Compare Sums and Differences of Fractions

✎ **Evaluate and compare. Write < or > or =.**

1) $\frac{1}{4} + \frac{3}{4} \square \frac{1}{4}$

2) $\frac{1}{3} + \frac{2}{3} \square 1$

3) $\frac{1}{5} + \frac{1}{5} \square \frac{2}{5}$

4) $\frac{1}{7} + \frac{2}{7} \square \frac{1}{7}$

5) $\frac{3}{8} + \frac{5}{8} \square \frac{1}{2}$

6) $\frac{7}{8} - \frac{3}{8} \square \frac{6}{8}$

7) $\frac{7}{12} + \frac{2}{12} \square \frac{7}{12}$

8) $\frac{7}{10} - \frac{5}{10} \square \frac{9}{10}$

9) $\frac{10}{15} - \frac{6}{15} \square \frac{3}{15}$

10) $\frac{3}{9} + \frac{1}{9} \square \frac{1}{9}$

11) $\frac{10}{12} + \frac{1}{12} \square \frac{9}{12}$

12) $\frac{17}{16} - \frac{3}{16} \square \frac{17}{16}$

13) $\frac{11}{17} + \frac{6}{17} \square \frac{18}{17}$

14) $\frac{15}{11} - \frac{5}{11} \square \frac{12}{11}$

15) $\frac{28}{32} - \frac{15}{32} \square \frac{25}{32}$

16) $\frac{25}{40} + \frac{15}{40} \square \frac{17}{40}$

17) $\frac{25}{53} - \frac{3}{53} \square \frac{9}{53}$

18) $\frac{35}{37} - \frac{20}{37} \square \frac{30}{37}$

19) $\frac{2}{12} + \frac{5}{12} \square \frac{11}{12}$

20) $\frac{18}{43} + \frac{13}{43} \square \frac{35}{43}$

Add 3 or More Fractions

✎ **Add fractions.**

1) $\frac{1}{5} + \frac{1}{5} + \frac{3}{5} =$

2) $\frac{1}{4} + \frac{1}{4} + \frac{1}{4} =$

3) $\frac{1}{11} + \frac{3}{11} + \frac{2}{11} =$

4) $\frac{5}{12} + \frac{6}{12} + \frac{1}{12} =$

5) $\frac{2}{6} + \frac{3}{6} + \frac{1}{6} =$

6) $\frac{4}{11} + \frac{5}{11} + \frac{1}{11} =$

7) $\frac{1}{2} + \frac{1}{2} + \frac{1}{2} =$

8) $\frac{6}{17} + \frac{5}{17} + \frac{4}{17} =$

9) $\frac{3}{15} + \frac{2}{15} + \frac{3}{15} =$

10) $\frac{4}{12} + \frac{2}{12} + \frac{1}{12} =$

11) $\frac{3}{25} + \frac{7}{25} + \frac{4}{25} =$

12) $\frac{7}{31} + \frac{15}{31} + \frac{4}{31} =$

13) $\frac{8}{40} + \frac{4}{40} + \frac{3}{40} =$

14) $\frac{2}{19} + \frac{3}{19} + \frac{7}{19} =$

15) $\frac{5}{42} + \frac{5}{42} + \frac{5}{42} =$

16) $\frac{13}{60} + \frac{9}{60} + \frac{12}{60} =$

17) $\frac{8}{17} + \frac{5}{17} + \frac{5}{17} =$

18) $\frac{3}{51} + \frac{5}{51} + \frac{6}{51} =$

19) $\frac{9}{38} + \frac{3}{38} + \frac{7}{38} =$

20) $\frac{7}{92} + \frac{15}{92} + \frac{14}{92} =$

21) $\frac{17}{43} + \frac{12}{43} + \frac{5}{43} =$

22) $\frac{22}{70} + \frac{15}{70} + \frac{13}{70} =$

Answers of Worksheets – Chapter 4

Fractions of a number

1) 40
2) 80
3) 36
4) 6
5) 2
6) 45
7) 7
8) 30
9) 3
10) 12
11) 42
12) 6
13) 10
14) 6
15) 93
16) 20
17) 420
18) 80
19) 24
20) 2

Order fractions

1) $\frac{4}{3}, \frac{1}{4}, \frac{1}{8}$
2) $\frac{14}{5}, \frac{1}{5}, \frac{1}{8}$
3) $\frac{5}{4}, \frac{6}{10}, \frac{1}{10}$
4) $\frac{3}{2}, \frac{3}{4}, \frac{2}{7}$
5) $\frac{3}{4}, \frac{1}{3}, \frac{3}{10}$
6) $\frac{3}{4}, \frac{1}{2}, \frac{1}{4}$
7) $\frac{7}{10}, \frac{1}{2}, \frac{1}{10}$
8) $\frac{9}{4}, \frac{5}{4}, \frac{3}{4}$
9) $\frac{5}{7}, \frac{3}{7}, \frac{2}{7}$
10) $\frac{8}{5}, \frac{5}{8}, \frac{1}{8}$
11) $\frac{5}{4}, \frac{7}{4}, \frac{8}{4}$
12) $\frac{4}{7}, \frac{5}{7}, \frac{8}{7}$
13) $\frac{1}{6}, \frac{3}{6}, \frac{5}{6}$
14) $\frac{1}{5}, \frac{2}{5}, \frac{3}{5}$
15) $\frac{1}{6}, \frac{3}{6}, \frac{5}{6}$
16) $\frac{3}{4}, \frac{5}{3}, \frac{8}{3}$
17) $\frac{2}{5}, \frac{8}{5}, \frac{11}{5}$
18) $\frac{11}{3}, \frac{17}{3}, \frac{22}{3}$

Simplifying Fractions

1) $\frac{1}{2}$
2) $\frac{4}{5}$
3) $\frac{3}{4}$
4) $\frac{1}{4}$
5) $\frac{3}{4}$
6) $\frac{2}{3}$
7) $\frac{4}{5}$
8) $\frac{1}{4}$
9) $\frac{1}{2}$
10) $\frac{2}{3}$
11) $\frac{1}{3}$
12) $\frac{3}{4}$
13) $\frac{9}{11}$
14) $\frac{9}{10}$
15) $\frac{3}{4}$
16) $\frac{5}{21}$
17) $\frac{5}{12}$
18) $\frac{3}{5}$
19) $\frac{8}{17}$
20) $\frac{3}{13}$
21) $\frac{1}{2}$
22) B
23) A
24) B

Common Core Exercise Book – Grade 3

Improper Fraction

1) $\frac{1}{3}$
2) $\frac{21}{5}$
3) $\frac{4}{5}$
4) $\frac{1}{2}$
5) $\frac{3}{7}$
6) $\frac{7}{4}$
7) $\frac{14}{5}$
8) $\frac{13}{4}$
9) $\frac{11}{7}$
10) $\frac{5}{6}$
11) $1\frac{1}{4}$
12) $3\frac{1}{2}$
13) $1\frac{2}{5}$
14) $6\frac{1}{2}$
15) $1\frac{3}{4}$
16) $4\frac{1}{3}$
17) $2\frac{1}{3}$
18) $2\frac{2}{4}$
19) $1\frac{4}{8}$
20) $2\frac{4}{5}$

Comparing Fractions and Missing Denominator

1) $\frac{1}{3} = \frac{3}{9}$
2) $\frac{4}{24} < \frac{1}{4}$
3) $\frac{24}{48} > \frac{4}{16}$
4) $\frac{8}{10} = \frac{4}{5}$
5) $\frac{15}{45} = \frac{1}{3}$
6) $\frac{21}{28} < \frac{6}{5}$
7) $\frac{6}{18} = \frac{12}{36}$
8) $\frac{11}{22} < \frac{20}{22}$
9) $\frac{1}{4} = \frac{5}{20}$
10) $\frac{1}{4} = \frac{4}{16}$
11) $\frac{3}{9} = \frac{1}{3}$
12) $\frac{4}{12} = \frac{1}{3}$
13) $\frac{5}{12} = \frac{20}{48}$
14) $\frac{9}{72} = \frac{1}{8}$
15) $\frac{3}{16} = \frac{6}{32}$
16) $\frac{7}{10} = \frac{21}{30}$
17) $\frac{3}{18} = \frac{1}{6}$
18) $\frac{4}{32} = \frac{1}{8}$

Add Fractions

1) 1
2) 1
3) $\frac{9}{7}$
4) 5
5) $\frac{7}{11}$
6) $\frac{7}{8}$
7) $\frac{11}{5}$
8) $\frac{11}{13}$
9) $\frac{15}{17}$
10) $\frac{8}{9}$
11) $\frac{11}{14}$
12) $\frac{16}{21}$
13) 1
14) $\frac{7}{19}$
15) $\frac{10}{15}$
16) $\frac{27}{35}$
17) $\frac{17}{25}$
18) $\frac{14}{22}$
19) $\frac{31}{43}$
20) $\frac{23}{39}$
21) $\frac{36}{71}$
22) $\frac{42}{52}$

Common Core Exercise Book – Grade 3

Subtract Fractions

1) $\frac{1}{7}$
2) $\frac{1}{5}$
3) $\frac{1}{3}$
4) $\frac{4}{9}$
5) $\frac{1}{12}$
6) $\frac{2}{15}$
7) $\frac{1}{3}$
8) $\frac{4}{15}$
9) $\frac{13}{25}$
10) $\frac{1}{24}$
11) $\frac{3}{31}$
12) $\frac{1}{28}$
13) $\frac{3}{25}$
14) $\frac{9}{20}$
15) $\frac{13}{20}$
16) $\frac{1}{5}$
17) $\frac{4}{23}$
18) $\frac{5}{41}$
19) $\frac{1}{2}$
20) $\frac{2}{7}$
21) $\frac{1}{6}$
22) $\frac{1}{28}$

Add and Subtract Fractions

1) 1
2) 1
3) $\frac{6}{7}$
4) $\frac{7}{9}$
5) $\frac{11}{12}$
6) $\frac{1}{4}$
7) $\frac{5}{9}$
8) 1
9) $\frac{1}{12}$
10) $\frac{21}{23}$
11) $\frac{2}{7}$
12) $\frac{1}{8}$
13) $\frac{1}{11}$
14) $\frac{7}{18}$
15) $\frac{5}{19}$
16) $\frac{2}{13}$
17) $\frac{2}{33}$
18) $\frac{2}{55}$
19) $\frac{2}{61}$
20) $\frac{5}{82}$

Compare Sums and Differences of Fractions

1) $1 > \frac{1}{4}$
2) $1 = 1$
3) $\frac{2}{5} = \frac{2}{5}$
4) $\frac{3}{7} > \frac{1}{7}$
5) $1 > \frac{1}{2}$
6) $\frac{4}{8} < \frac{6}{8}$
7) $\frac{9}{12} > \frac{7}{12}$
8) $\frac{2}{10} < \frac{9}{10}$
9) $\frac{4}{15} > \frac{3}{15}$
10) $\frac{4}{9} > \frac{1}{9}$
11) $\frac{11}{12} > \frac{9}{12}$
12) $\frac{14}{16} < \frac{17}{16}$
13) $1 > \frac{18}{21}$
14) $\frac{10}{11} < \frac{12}{11}$
15) $\frac{16}{32} < \frac{25}{32}$
16) $1 > \frac{15}{40}$

Common Core Exercise Book – Grade 3

17) $\frac{22}{53} > \frac{9}{53}$

18) $\frac{15}{37} < \frac{30}{37}$

19) $\frac{7}{12} > \frac{11}{12}$

20) $\frac{31}{43} > \frac{35}{43}$

Add 3 or More Fractions

1) 1

2) $\frac{3}{4}$

3) $\frac{6}{11}$

4) 1

5) 1

6) $\frac{10}{11}$

7) $\frac{3}{2}$

8) $\frac{15}{17}$

9) $\frac{8}{15}$

10) $\frac{7}{12}$

11) $\frac{14}{25}$

12) $\frac{26}{31}$

13) $\frac{15}{40}$

14) $\frac{13}{19}$

15) $\frac{15}{42}$

16) $\frac{29}{60}$

17) $\frac{18}{17}$

18) $\frac{14}{51}$

19) $\frac{19}{38}$

20) $\frac{36}{92}$

21) $\frac{34}{43}$

22) $\frac{50}{70}$

Common Core Exercise Book – Grade 3

Chapter 5:
Time and Money

Topics that you'll practice in this chapter:

- ✓ Read Clocks
- ✓ Telling Time
- ✓ Digital Clock
- ✓ Measurement – Time
- ✓ Add Money Amounts
- ✓ Subtract Money Amounts
- ✓ Money: Word Problems

Read Clocks

✎ **Write the time below each clock.**

1) _____ 2) _____ 3) _____

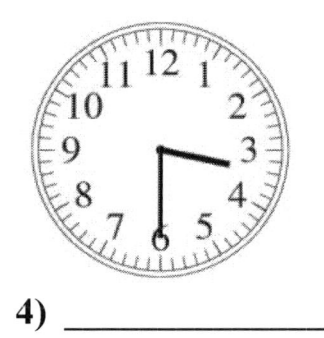

4) _____ 5) _____ 6) _____

✎ **How much time has passed?**

7) From 2:15 AM to 5:35 AM: _____ hours and _____ minutes.

8) From 2:50 AM to 8:05 AM: _____ hours and _____ minutes.

9) It's 6:30 P.M. What time was 2 hours ago? _____ O'clock

Digital Clock

✎ **What time is it? Write the time in words in front of each.**

1) 4 : 30 _____

2) 2 : 14 _____

3) 6 : 42 _____

4) 8 : 20 _____

5) 11 : 10 _____

6) 12 : 15 _____

7) 3 : 25 _____

8) 6 : 17 _____

9) 8 : 56 _____

10) 1 : 12 _____

11) 9 : 02 _____

12) 7 : 34 _____

Measurement – Time

✍ How much time has passed?

1) 2:50 AM to 5:35 AM: _____ hours and _____ minutes.

2) 4:15 AM to 7:10 AM: _____ hours and _____ minutes.

3) 9:00 AM. to 11:25 AM. = _____ hour(s) and _____ minutes.

4) 8:15 PM to 9:45 PM. = _____ hour(s) and _____ minutes

5) 7:15 A.M. to 7:45 A.M. = _____ minutes

6) 10:05 A.M. to 10:30 A.M. = _____ minutes

7) There are _____ second in 17 minutes.

8) There are _____ second in 16 minutes.

9) There are _____ second in 29 minutes.

10) There are _____ minutes in 25 hours.

11) There are _____ minutes in 40 hours.

12) There are _____ minutes in 18 hours.

Add Money Amounts

✎ **Add.**

1) $412 $952 $460
   +$126   +$310   +$315
   -----   -----   -----

2) $326 $590 $815
   +$230   +$425   +$165
   -----   -----   -----

3) $211 $568 $762
   +$415   +$245   +$315
   -----   -----   -----

4) $356 $321 $705
   +$425   +$129   +$381
   -----   -----   -----

5) $726 $365 $536
   +$125   +$132   +$215
   -----   -----   -----

6) $396 $312 $806
   +$264   +$196   +$237
   -----   -----   -----

7) $265 $685 $745
   +$345   +$264   +$362
   -----   -----   -----

Subtract Money Amounts

✎ **Subtract.**

1) $645 − $154 $862 − $330 $735 − $652

2) $468 − $136 $496 − $369 $963 − $654

3) $862 − $430 $550 − $150 $634 − $456

4) $362 − $129 $432 − $385 $469 − $251

5) $462 − $298 $563 − $462 $786 − $405

6) $836 − $563 $925 − $735 $993 − $746

7) Mia had $150. She bought some game tickets for $80. How much did she have left?

Money: Word Problems

✎ **Solve.**

1) How many boxes of envelopes can you buy with $12 if one box costs $2?

2) After paying $8.35 for a salad, Ella has $42.25. How much money did she have before buying the salad?

3) How many packages of diapers can you buy with $60 if one package costs $6?

4) Last week James ran 40 miles more than Mason. James ran 86 miles. How many miles did Mason run?

5) Last Friday Joy had $36.55. Over the weekend he received some money for cleaning the attic. He now has $64. How much money did he receive?

6) After paying $12.15 for a sandwich, Amelia has $25.75. How much money did she have before buying the sandwich?

Answers of Worksheets – Chapter 5

Read clocks

1) 1
2) 4 :45
3) 8
4) 3 :30
5) 10 :15
6) 8 :35
7) 3 hours and 20 minutes
8) 5 hours and 15 minutes
9) 4 : 30 PM

Digital Clock

1) It's four thirty.
2) It's two Fourteen.
3) It's six forty–two.
4) It's eight twenty.
5) It's eleven ten.
6) It's Twelve Fifteen.
7) It's three Twenty–five.
8) It's six seventeen.
9) It's eight fifty-six.
10) It's one Twelve.
11) It's nine two.
12) It's seven thirty-four.

Measurement – Time

1) 2 :45
2) 2 :55
3) 2 :25
4) 1 :30
5) 30 minutes
6) 25 minutes
7) 1,020
8) 960
9) 1,740
10) 1,500
11) 2,400
12) 1,080

Add Money Amounts

1) 538, 1,262, 775
2) 556, 1,015, 980
3) 626, 813, 1,077
4) 781, 450, 1,086
5) 851, 497, 751
6) 660, 508, 1,043
7) 610, 949, 1,107

Subtract Money Amounts

1) 491–532–83
2) 332–127–309
3) 432–400–178
4) 233, 47, 218
5) 164, 101, 318
6) 273, 190, 247
7) 70

Money: word problem

1) 6
2) $50.6
3) 10
4) 46
5) 27.45
6) 37.9

Chapter 6: Measurement

Topics that you'll learn in this chapter:

- ✓ Reference Measurement
- ✓ Metric Length
- ✓ Customary Length
- ✓ Metric Capacity
- ✓ Customary Capacity
- ✓ Metric Weight and Mass
- ✓ Customary Weight and Mass
- ✓ Time
- ✓ Add Money Amounts
- ✓ Subtract Money Amounts
- ✓ Money: Word Problems

Metric Length Measurement

📝 **Convert to the units.**

1) 40 mm = _____ cm

2) 4 m = _____ mm

3) 6 m = _____ cm

4) 5 km = _____ m

5) 3,000 mm = _____ m

6) 800 cm = _____ m

7) 20 m = _____ cm

8) 2,000 mm = _____ cm

9) 6,000 mm = _____ m

10) 8 km = _____ mm

11) 20 km = _____ m

12) 70 m = _____ cm

13) 8,000 m = _____ km

14) 2,000 m = _____ km

Customary Length Measurement

📝 **Convert to the units.**

1) 10 ft = _____ in

2) 4 ft = _____ in

3) 8 yd = _____ ft

4) 16 yd = _____ ft

5) 4 yd = _____ in

6) 14 in = _____ ft

7) 432 in = _____ yd

8) 216 in = _____ yd

9) 57 yd = _____ in

10) 34 yd = _____ in

11) 72 ft = _____ yd

12) 240 ft = _____ yd

13) 36 in = _____ ft

14) 56 yd = _____ feet

15) 30 in = _____ ft

16) 60 in = _____ ft

Metric Capacity Measurement

✎ **Convert the following measurements.**

1) 40 l = _____ ml

2) 7 l = _____ ml

3) 60 l = _____ ml

4) 33 l = _____ ml

5) 42 l = _____ ml

6) 18 l = _____ ml

7) 40,000 l = _____ l

8) 35,000m ml = _____ l

9) 62,000 ml = _____ l

10) 5000 ml = _____ l

11) 7000 ml = _____ l

12) 40, 000 ml = _____ l

Customary Capacity Measurement

✎ **Convert the following measurements.**

1) 36 gal = _____ qt.

2) 32 gal = _____ pt.

3) 82 gal = _____ c.

4) 7 pt. = _____ c

5) 96 qt = _____ pt.

6) 15 qt = _____ c

7) 22 pt. = _____ c

8) 72 c = _____ gal

9) 288 pt. = _____ gal

10) 248 qt = _____ gal

11) 324 pt. = _____ qt

12) 100 c = _____ qt

13) 138 c = _____ pt.

14) 244 qt = _____ gal

15) 184 pt. = _____ qt

16) 36 gal = _____ pt.

17) 62 qt = _____ c

18) 400 c = _____ gal

Metric Weight and Mass Measurement

✍ Convert.

1) 60 kg = _____ g
2) 66 kg = _____ g
3) 300 kg = _____ g
4) 60 kg = _____ g
5) 32 kg = _____ g
6) 60 kg = _____ g
7) 28 kg = _____ g

8) 36,000 g = _____ kg
9) 130,000 g = _____ kg
10) 900,000 g = _____ kg
11) 30,000 g = _____ kg
12) 20,000 g = _____ kg
13) 450,000 g = _____ kg
14) 200,000 g = _____ kg

Customary Weight and Mass Measurement

✍ Convert.

1) 8,000 lb. = _____ T
2) 20,000 lb. = _____ T
3) 4,000 lb. = _____ T
4) 32,000 lb. = _____ T
5) 30 lb. = _____ oz
6) 52 lb. = _____ oz
7) 70 lb. = _____ oz
8) 2 T = _____ lb.

9) 9 T = _____ lb.
10) 12 T = _____ lb.
11) 50 T = _____ lb.
12) 18 T = _____ oz
13) 7 T = _____ oz
14) 6 T = _____ oz
15) 19 T = _____ lb
16) 36 T = _____ lb.

Answers of Worksheets – Chapter 6

Metric length

1) 4 cm
2) 4000 mm
3) 600 cm
4) 5000 m
5) 3 m
6) 8 m
7) 2000 cm
8) 20 cm
9) 6 m
10) 8,000,000 mm
11) 20,000 m
12) 7,000 cm
13) 8 km
14) 2 km

Customary Length

1) 120
2) 48
3) 24
4) 48
5) 144
6) 7
7) 12
8) 6
9) 2,052
10) 1,224
11) 24
12) 80
13) 3
14) 168
15) 2.5
16) 5

Metric Capacity

1) 40,000 ml
2) 7,000 ml
3) 60,000 ml
4) 33,000 ml
5) 42,000 ml
6) 18,000 ml
7) 40 ml
8) 35 ml
9) 62 ml
10) 5L
11) 7L
12) 40 L

Customary Capacity

1) 144 qt
2) 256 pt.
3) 1312 c
4) 14 c
5) 192 pt.
6) 60 c
7) 44 c
8) 4.5 gal
9) 36 gal
10) 62 gal
11) 162 qt
12) 25 qt
13) 69 pt.
14) 61 gal
15) 92 qt
16) 288 pt.
17) 248 c
18) 25 gal

Metric Weight and Mass

1) 60,000 g
2) 66,000 g
3) 300,000 g
4) 60,000 g
5) 32,000 g
6) 60,000 g
7) 28,000 g
8) 36 kg
9) 130 kg
10) 900 kg
11) 30 kg
12) 20 kg

13) 450 kg   14) 200 kg

Customary Weight and Mass

1) 4 T   7) 1,120 oz   13) 224,000 oz

2) 10 T   8) 4,000 lb.   14) 192,000 oz

3) 2 T   9) 18,000 lb.   15) 38,000 lb

4) 16 T   10) 24,000 lb.   16) 72,000 lb

5) 480 oz   11) 100,000 lb.

6) 832 oz   12) 576,000 oz

Chapter 7: Symmetry

Topics that you'll practice in this chapter:

- ✓ Line Segments

- ✓ Identify Lines of Symmetry

- ✓ Count Lines of Symmetry

- ✓ Parallel, Perpendicular and Intersecting Lines

Line Segments

✎ **Write each as a line, ray or line segment.**

1)

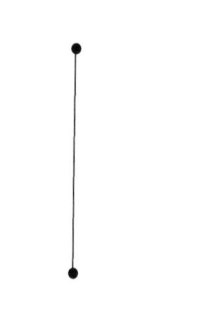

2)

3)

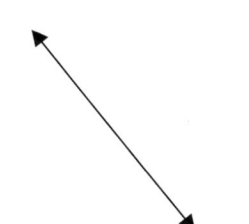

4)

5)

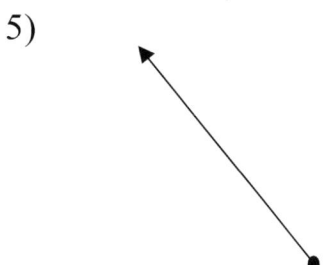

6)

7)

8)

Parallel, Perpendicular and Intersecting Lines

✎ State whether the given pair of lines are parallel, perpendicular, or intersecting.

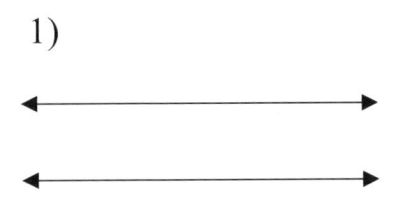

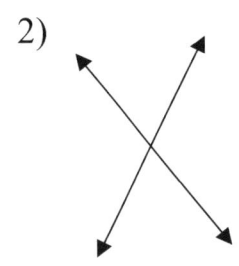

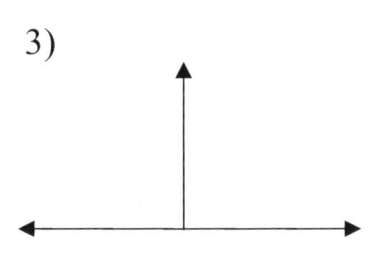

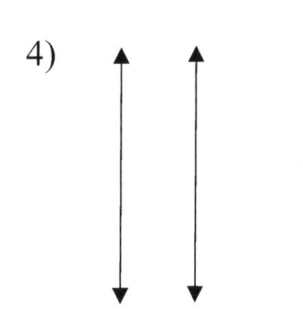

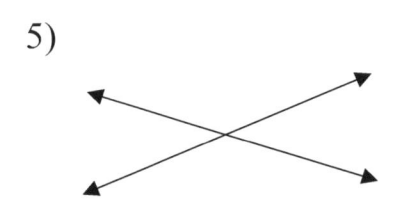

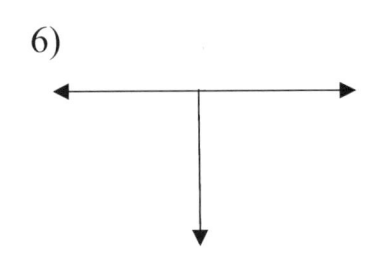

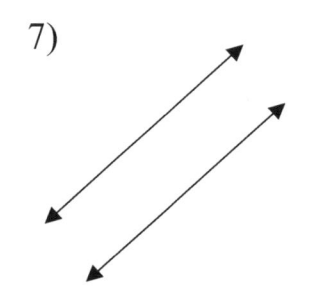

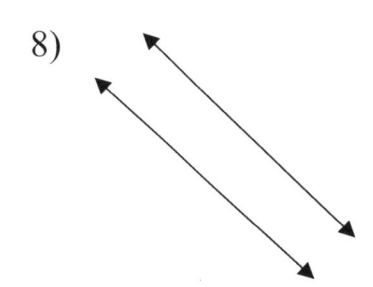

1)

2)

3)

4)

5)

6)

7)

8)

Identify Lines of Symmetry

✎ **Tell whether the line on each shape a line of symmetry is.**

1)

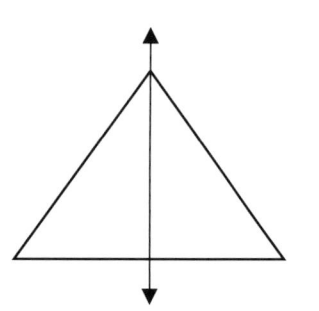

2)

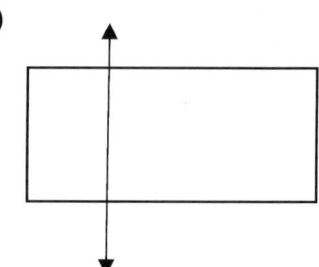

3)

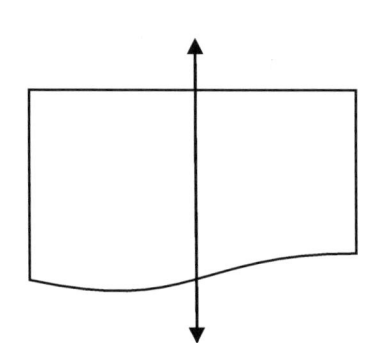

4)

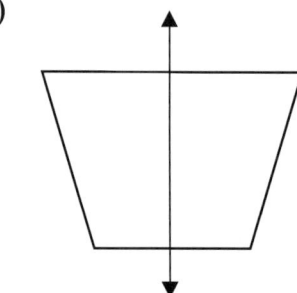

5)

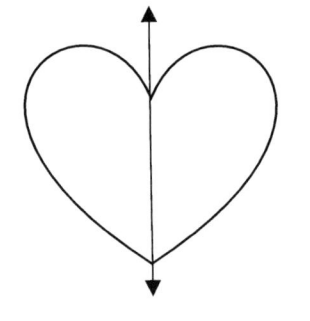

6)

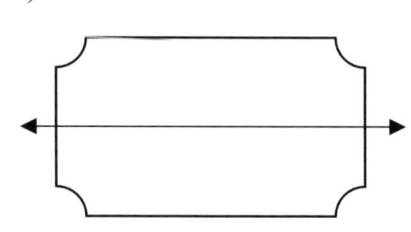

7)

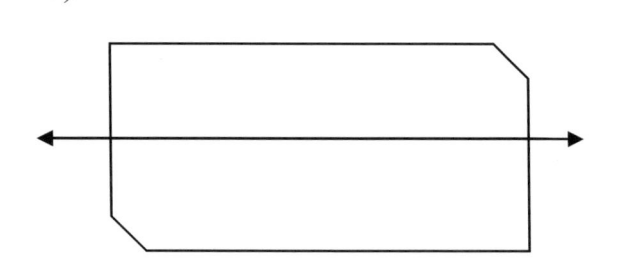

8)

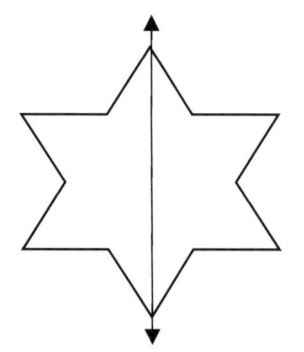

Lines of Symmetry

✎ **Draw lines of symmetry on each shape. Count and write the lines of symmetry you see.**

1)

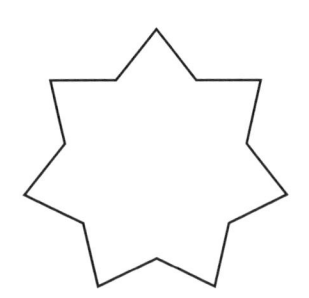

2)

3)

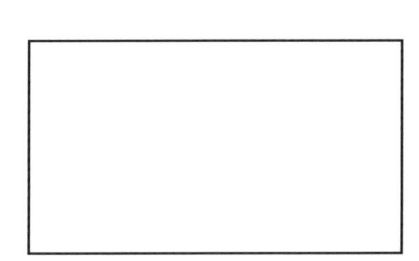

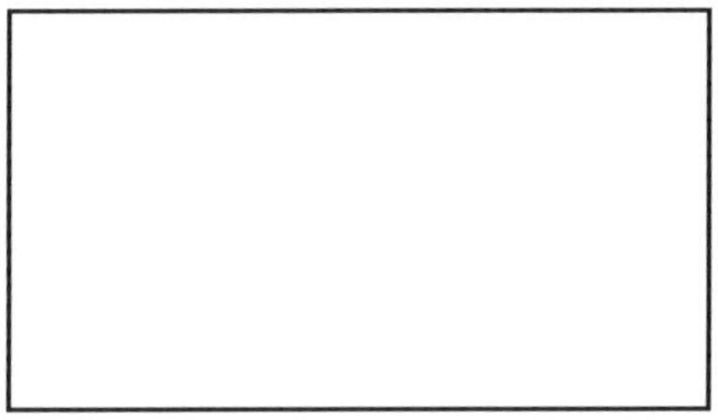

4)

5)

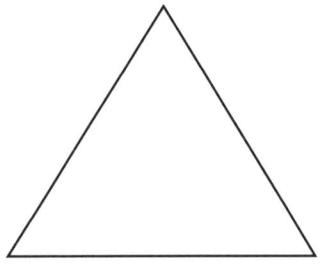

6)

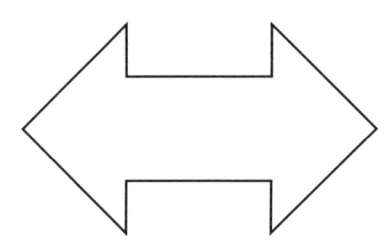

7)

8)

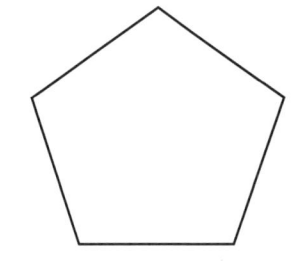

Answers of Worksheets – Chapter 7

Line Segments

1) Line segment
2) Ray
3) Line
4) Line segment
5) Ray
6) Line
7) Line
8) Line segment

Parallel, Perpendicular and Intersecting Lines

1) Parallel
2) Intersection
3) Perpendicular
4) Parallel
5) Intersection
6) Perpendicular
7) Parallel
8) Parallel

Identify lines of symmetry

1) yes
2) no
3) no
4) yes
5) yes
6) yes
7) no
8) yes

lines of symmetry

1)

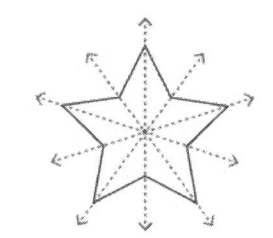

2)

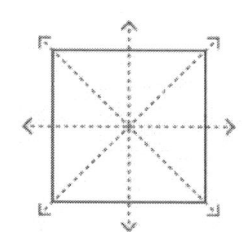

3)

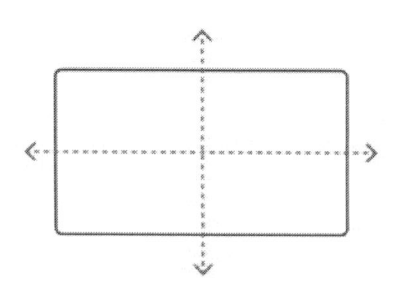

4)

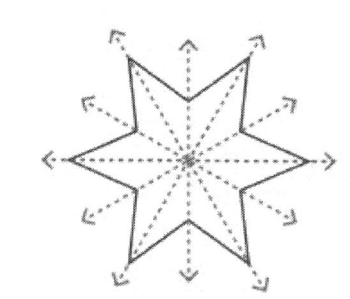

5)
6)
7)
8)

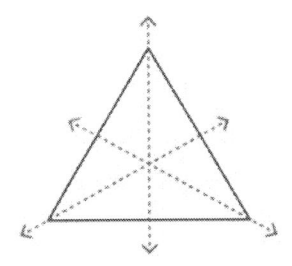

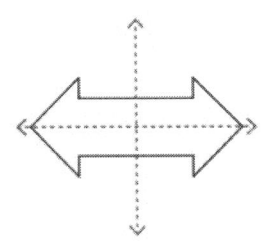

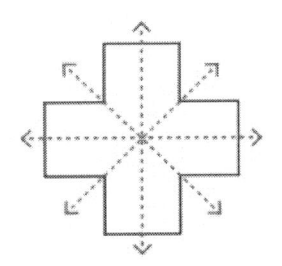

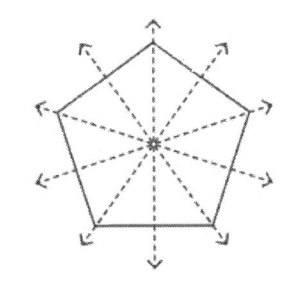

Chapter 8: Geometric

Topics that you'll practice in this chapter:

- ✓ Identifying Angles: Acute, Right, Obtuse, and Straight Angles
- ✓ Polygon Names
- ✓ Triangles
- ✓ Quadrilaterals and Rectangles
- ✓ Perimeter: Find the Missing Side Lengths
- ✓ Perimeter and Area of Squares
- ✓ Perimeter and Area of rectangles
- ✓ Area and Perimeter: Word Problems
- ✓ Area of Squares and Rectangles

Identifying Angles

✍ **Write the name of the angles(Acute, Right, Obtuse, and Straight Angles).**

1)

2)

3)

4)

5)

6)

7)

8)

Polygon Names

✎ **Write name of polygons.**

1)

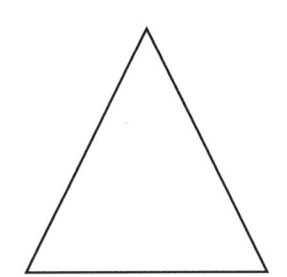

2)

3)

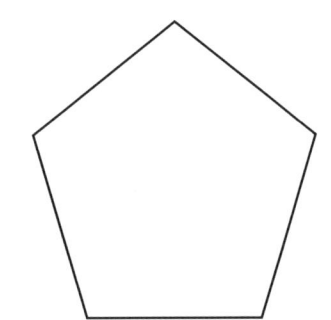

4)

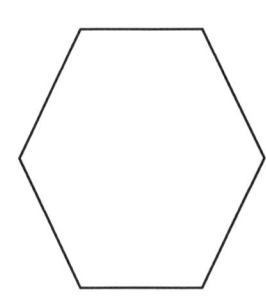

5)

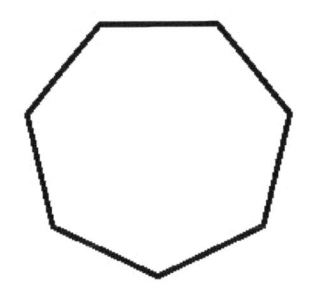

6)

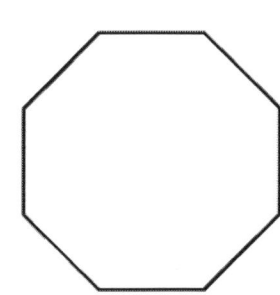

7)

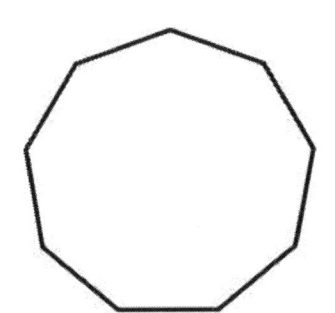

8)

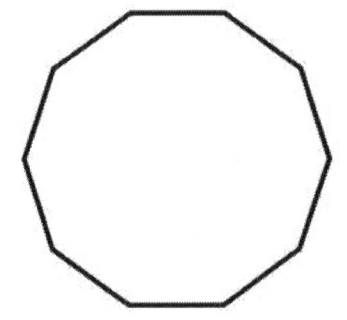

Common Core Exercise Book – Grade 3

Triangles

✎ **Classify the triangles by their sides and angles.**

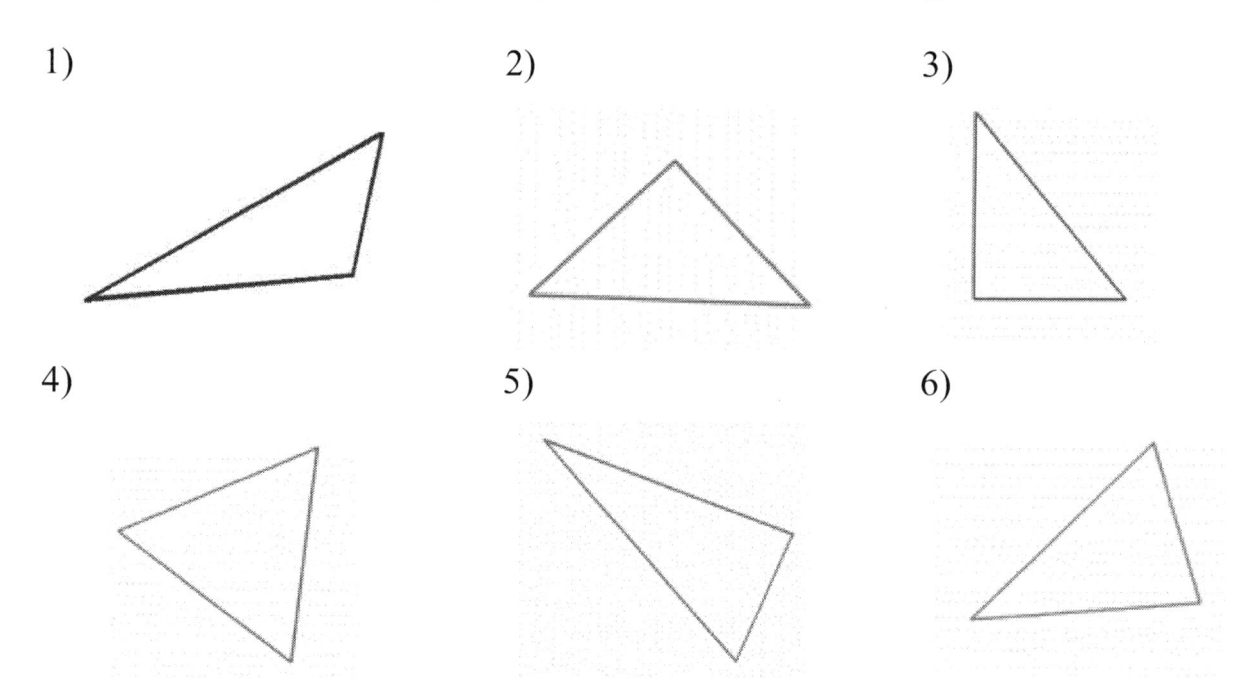

1)   2)   3)

4)   5)   6)

✎ **Find the measure of the unknown angle in each triangle.**

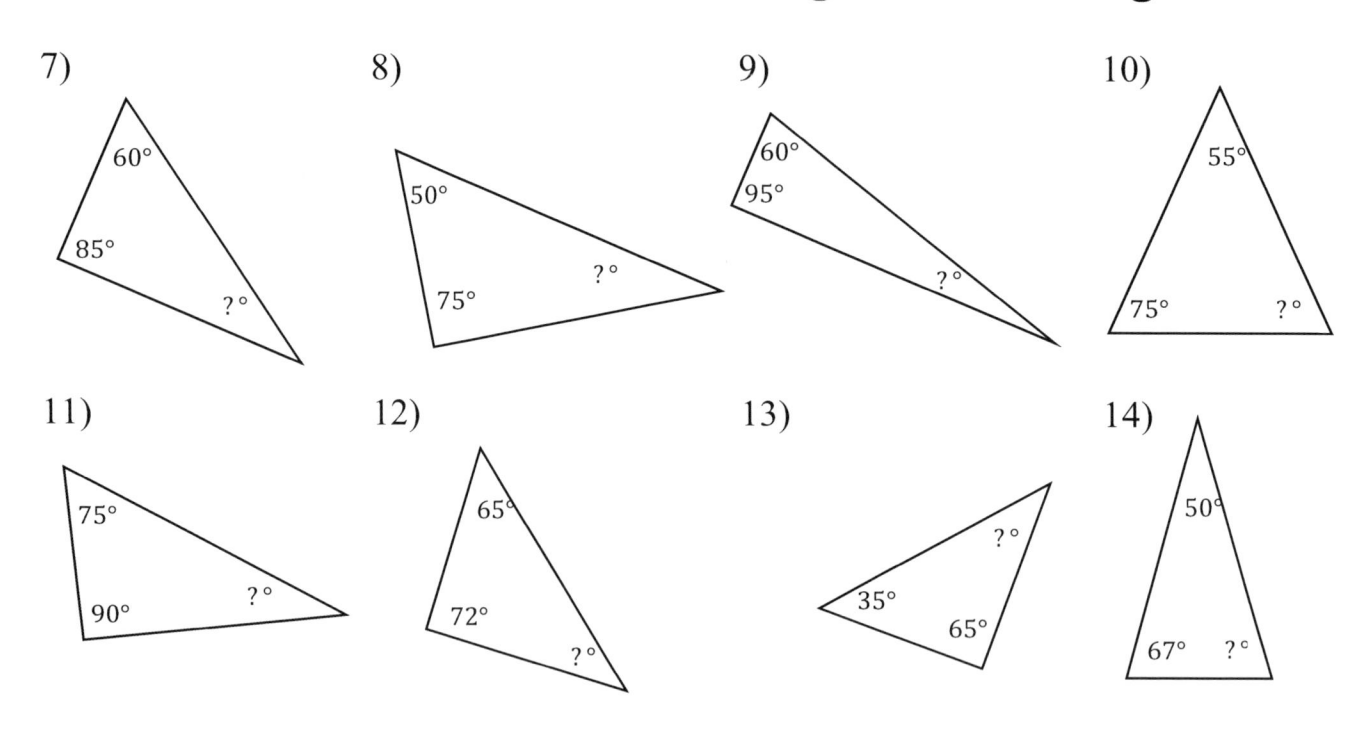

7) 60°, 85°, ?°

8) 50°, 75°, ?°

9) 60°, 95°, ?°

10) 55°, 75°, ?°

11) 75°, 90°, ?°

12) 65°, 72°, ?°

13) ?°, 35°, 65°

14) 50°, 67°, ?°

84

Quadrilaterals and Rectangles

✎ **Write the name of quadrilaterals.**

1)

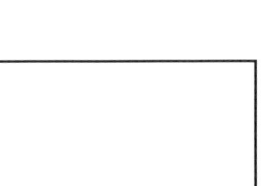

2)

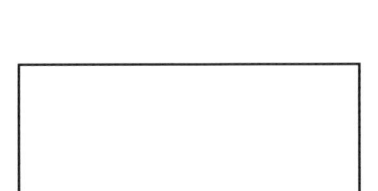

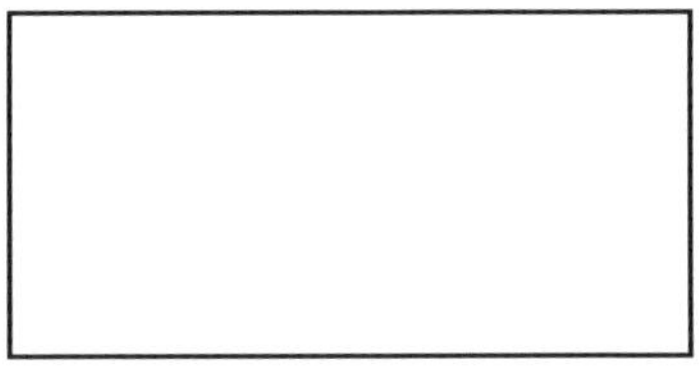

3)

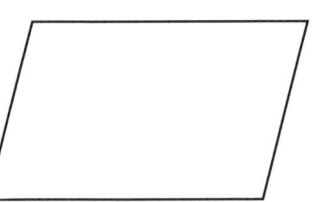

4)

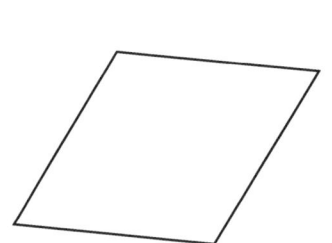

5)

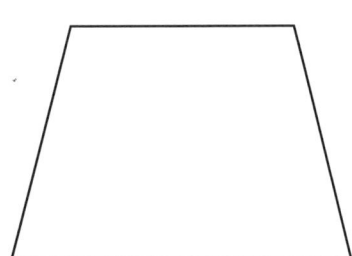

6)

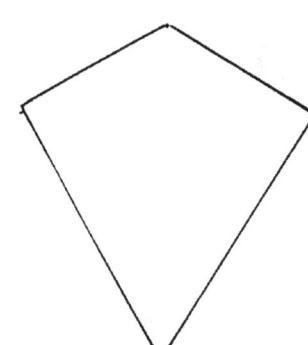

✎ **Solve.**

7) A rectangle has _____ sides and _____ angles.

8) Draw a rectangle that is 5 centimeters long and 4 centimeters wide. What is the perimeter?

9) Draw a rectangle 6 cm long and 3 cm wide.

10) Draw a rectangle whose length is 5cm and whose width is 3 cm. What is the perimeter of the rectangle?

11) What is the perimeter of the rectangle?

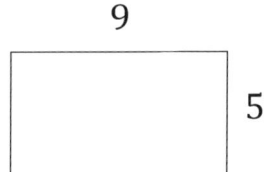

Perimeter and Area of Squares

✎ **Find perimeter and area of squares.**

1) A: _____, P: _____

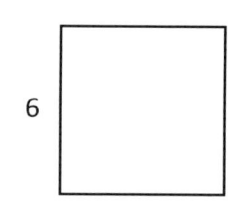

   6

2) A: _____, P: _____

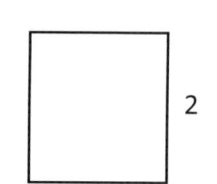

   2

3) A: _____, P: _____

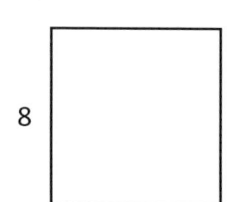

   8

4) A: _____, P: _____

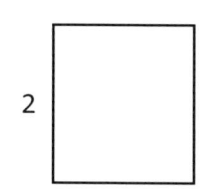

   2

5) A: _____, P: _____

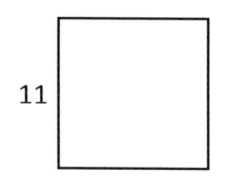

   11

6) A: _____, P: _____

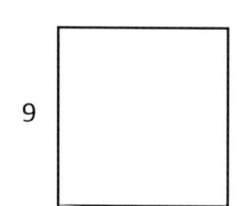

   9

7) A: _____, P: _____

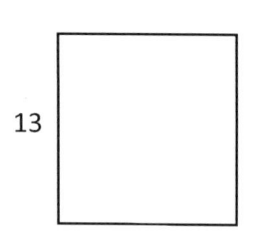

   13

8) A: _____, P: _____

   15

9) A: _____, P: _____

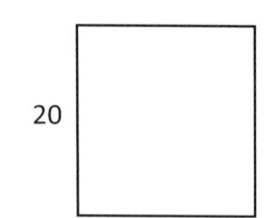

   20

10) A: _____, P: _____

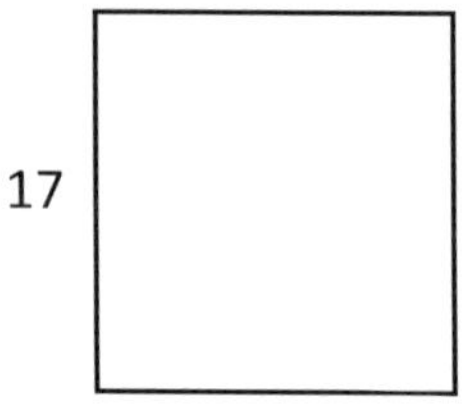

   17

Perimeter and Area of rectangles

✍ **Find perimeter and area of rectangles.**

1) A: ___, P: ___

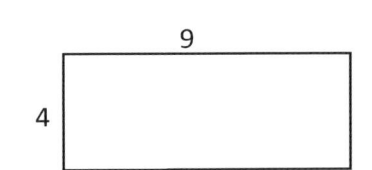

   9 × 4

2) A: ___, P: ___

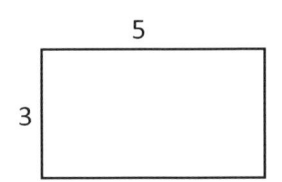

   5 × 3

3) A: ___, P: ___

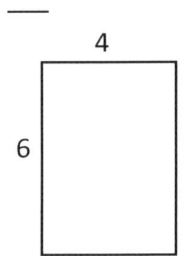

   4 × 6

4) A: ___, P: ___

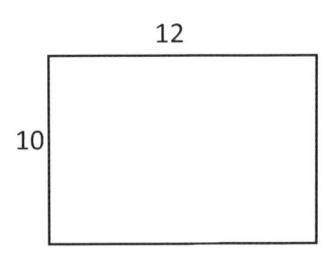

   12 × 10

5) A: ___, P: ___

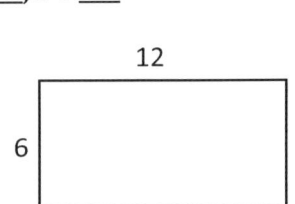

   12 × 6

6) A: ___, P: ___

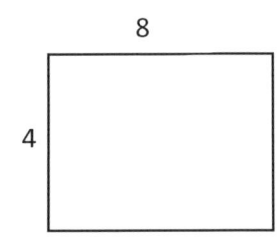

   8 × 4

7) A: ___, P: ___

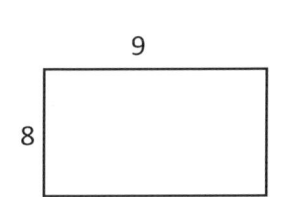

   9 × 8

8) A: ___, P: ___

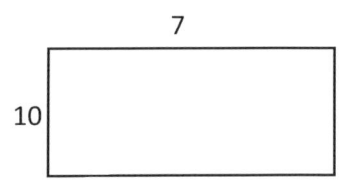

   7 × 10

9) A: ___, P: ___

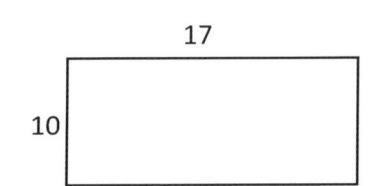

   17 × 10

10) A: ___, P: ___

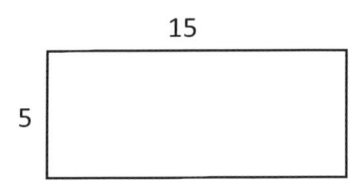

   15 × 5

Word Problem

✎ Find the area of each.

1)
 9 yd
 12 yd   12 yd
 9 yd

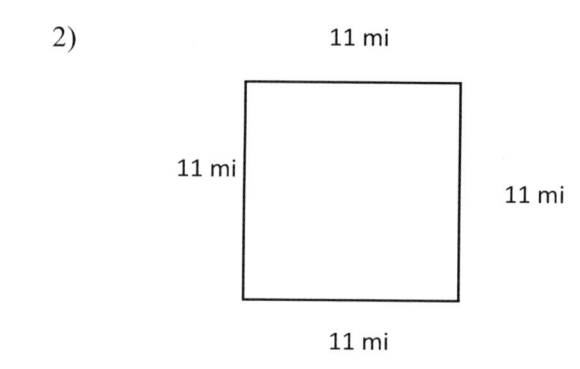

2)
 11 mi
 11 mi   11 mi
 11 mi

✎ Solve.

3) The area of a rectangle is 63 square meters. The width is 7 meters. What is the length of the rectangle?

4) A square has an area of 81 square feet. What is the perimeter of the square?

5) Ava built a rectangular vegetable garden that is 7 feet long and has an area of 42 square feet. What is the perimeter of Ava's vegetable garden?

6) A square has a perimeter of 48 millimeters. What is the area of the square?

7) The perimeter of David's square backyard is 80 meters. What is the area of David's backyard?

8) The area of a rectangle is 45 square inches. The length is 5 inches. What is the perimeter of the rectangle?

Answers of Worksheets – Chapter 8

Identifying Angles

1) Obtuse
2) Acute
3) Right
4) Acute
5) Straight
6) Obtuse
7) Obtuse
8) Acute

Polygon Names

1) Triangle
2) Quadrilateral
3) Pentagon
4) Hexagon
5) Heptagon
6) Octagon
7) Nonagon
8) Decagon

Triangles

1) Scalene, obtuse
2) Isosceles, right
3) Scalene, right
4) Equilateral, acute
5) Scalene, acute
6) Scalene, acute
7) 35°
8) 55°
9) 25°
10) 50°
11) 15°
12) 43°
13) 80°
14) 63°

Quadrilaterals and Rectangles

1) Square
2) Rectangle
3) Parallelogram
4) Rhombus
5) Trapezoid
6) Kike
7) 4 - 4
8) 18
9) Use a rule to draw the rectangle
10) 16
11) 28

Perimeter and Area of Squares

1) A:36, P: 24
2) A: 4, P: 8
3) A: 64, P: 32
4) A: 4, P: 8
5) A: 121, P: 44
6) A: 81, P: 36
7) A: 169, P: 52
8) A: 225, P: 60
9) A: 400, P: 80
10) A: 289, P: 68

Perimeter and Area of rectangles

1) A: 36, P: 26
2) A: 15, P: 16
3) A: 24, P: 20
4) A: 120, P: 44
5) A: 72, P: 36
6) A: 32, P: 24
7) A: 72, P: 34
8) A: 70, P: 34
9) A: 170, P: 54
10) A: 75, P: 40

Word Problem

1) 108 yd^2
2) 121 mi^2
3) 9
4) 36
5) 26
6) 144
7) 400
8) 28

Chapter 9:
Patterns and Graphs

Topics that you'll practice in this chapter:

- ✓ Repeating pattern
- ✓ Growing Patterns
- ✓ Patterns: Numbers
- ✓ Bar Graph
- ✓ Tally and Pictographs
- ✓ Line Graphs

Common Core Exercise Book – Grade 3

Repeating Pattern

✎ **Circle the picture that comes next in each picture pattern.**

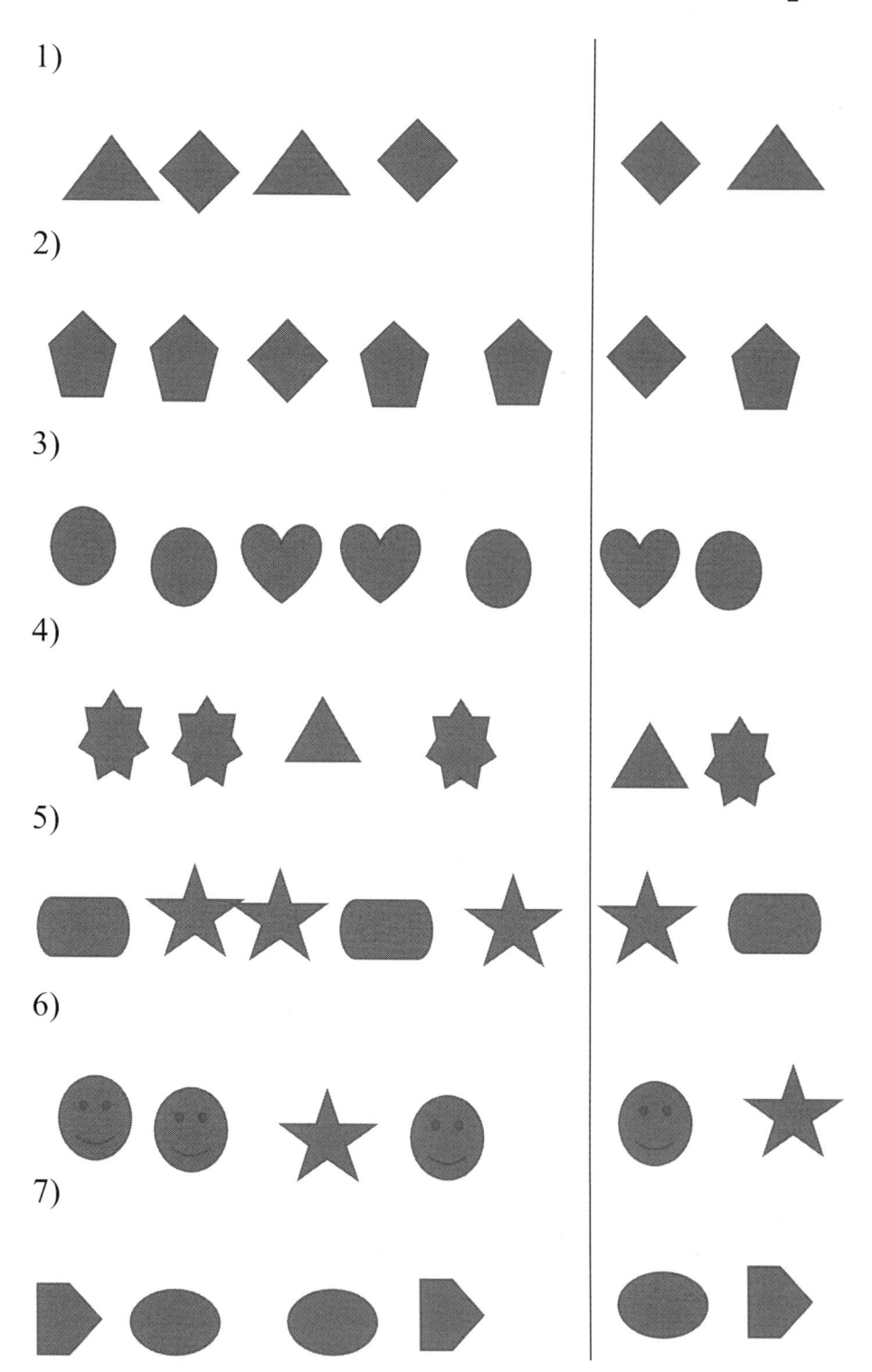

1)

2)

3)

4)

5)

6)

7)

Growing Patterns

✍ **Draw the picture that comes next in each growing pattern.**

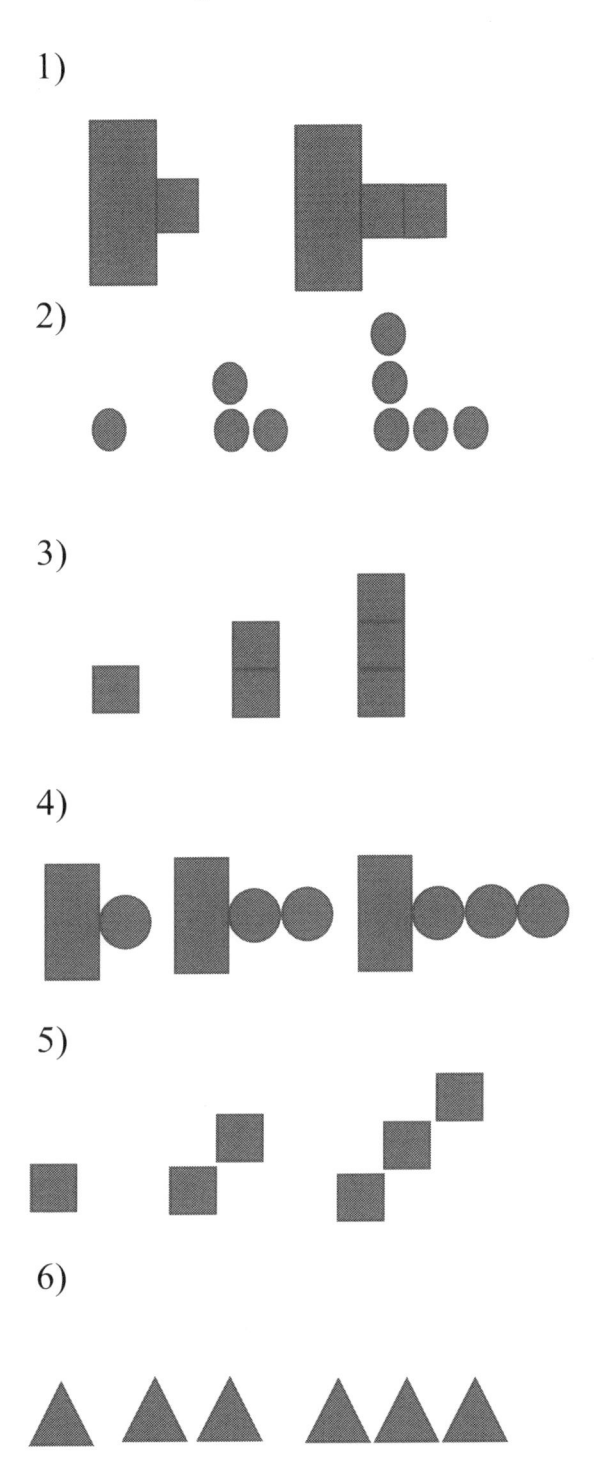

1)

2)

3)

4)

5)

6)

Patterns: Numbers

Write the numbers that come next.

1) 11, 14, 17, 20, ____, ____, ____, ____

2) 8, 16, 24, 32, ____, ____, ____, ____

3) 10, 25, 40, 55, ____, ____, ____, ____

4) 9, 18, 27, 36, ____, ____, ____, ____

5) 5, 10, 15, 20, 25, ____, ____, ____, ____

6) 60, 55, 45, 40, 35, ____, ____, ____, ____

7) 13, 25, 37, 49, ____, ____, ____, ____

Write the next three numbers in each counting sequence.

8) −30, −23, −16, ____, ____, ____, ____

9) 625, 605, 585, ____, ____, ____, ____

10) ____, ____, 58, 68, ____, 88

11) 25, 27, ____, ____, 33, ____

12) 18, 12, ____, ____, ____

13) 62, 55, ____, ____, ____

14) 43, 40, 37, ____, ____, ____

15) 62, 44, 26, ____, ____, ____

Bar Graph

✎ **Graph the given information as a bar graph.**

Day	Hot dogs sold
Monday	70
Tuesday	30
Wednesday	50
Thursday	10
Friday	60

Tally and Pictographs

✎ **Using the key, draw the pictograph to show the information.**

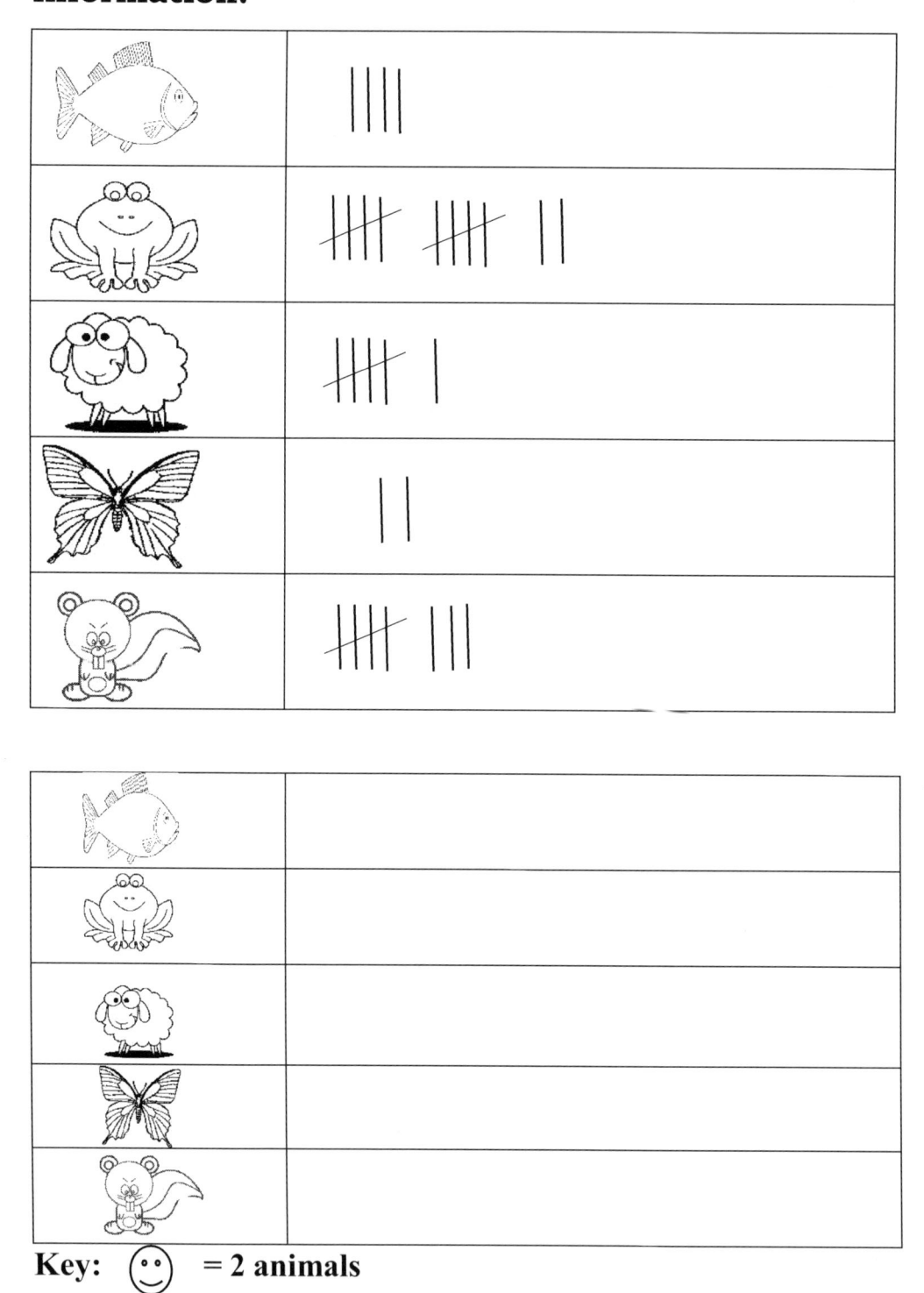

🐟	||||
🐸	|||| |||| ||
🐑	||||	
🦋	||
🐿️	|||| |||

🐟	
🐸	
🐑	
🦋	
🐿️	

Key: 😊 = 2 animals

Line Graphs

✍ David work as a salesman in a store. He records the number of phones sold in five days on a line graph. Use the graph to answer the questions.

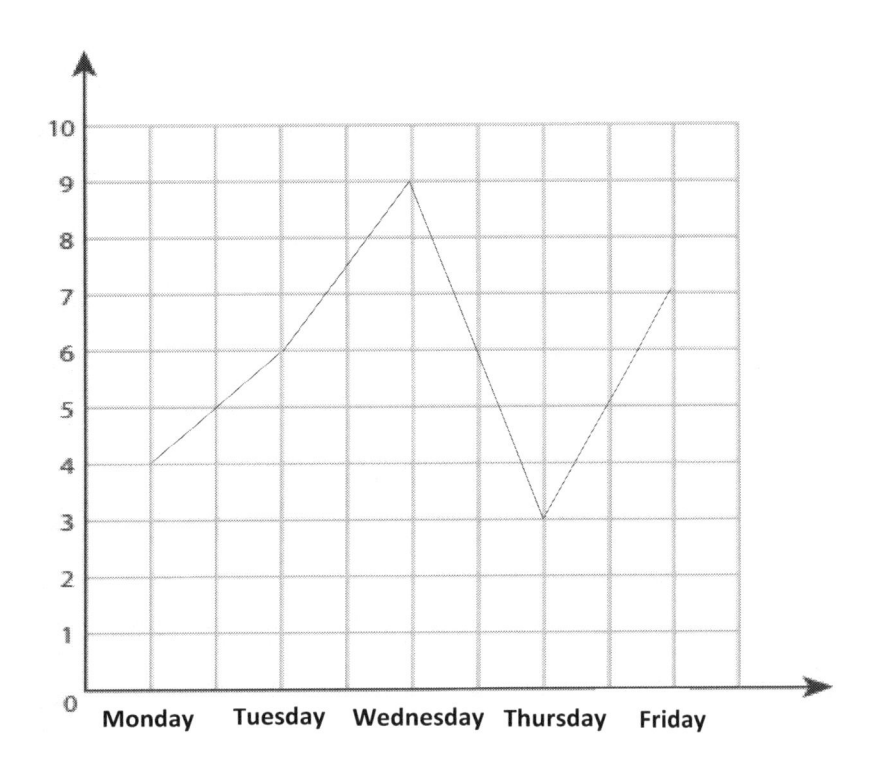

1) How many phones were sold on Tuesday?

2) Which day had the minimum sales of phones?

3) Which day had the maximum number of phones sold?

4) How many phones were sold in 5 days?

Answers of Worksheets – Chapter 9

Repeating pattern

1)  ▲
2) ◆
3) 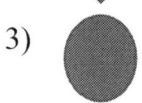 ●
4)  ✦

5)  ★
6) ☺
7) ⬭

Growing patterns

1) 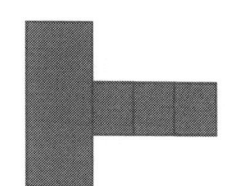
2) 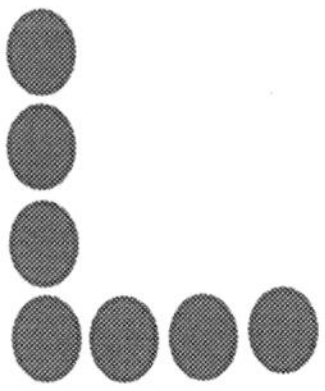
3)
4) 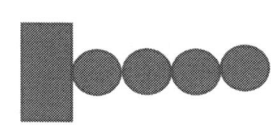
5) 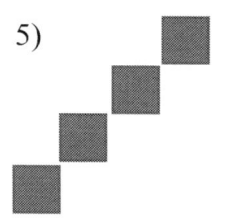
6)

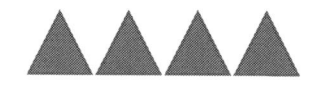

Common Core Exercise Book – Grade 3

Bar Graph

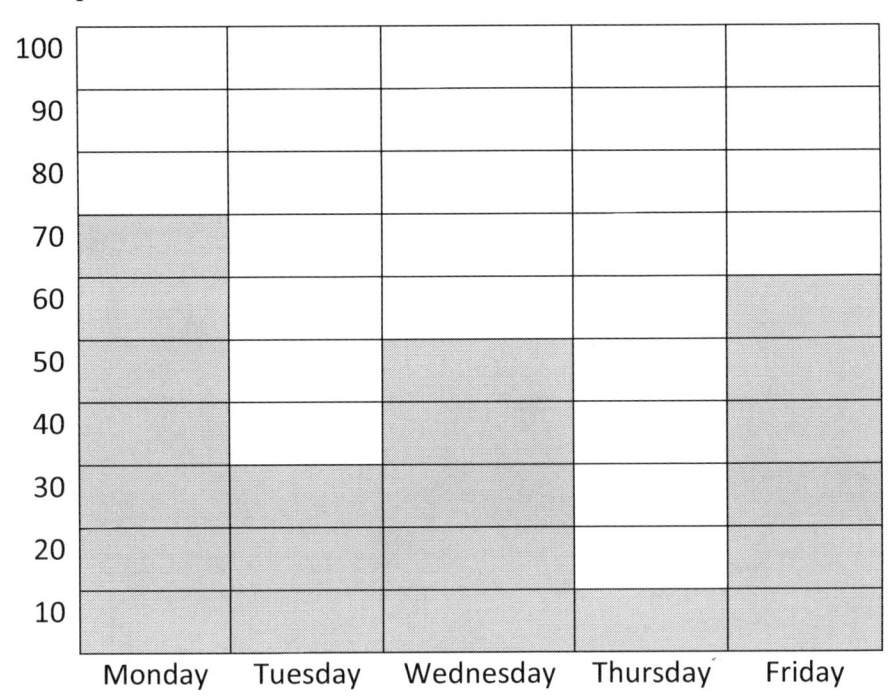

Tally and Pictographs

Line Graphs

1) 6
2) Thursday
3) Wednesday
4) 29

Common Core Math Practice Tests

Time to Test

Time to refine your skill with a practice examination

Take a REAL Common Core Mathematics test to simulate the test day experience. After you've finished, score your test using the answer key.

Before You Start

- You'll need a pencil and scratch papers to take the test.
- For this practice test, don't time yourself. Spend time as much as you need.
- It's okay to guess. You won't lose any points if you're wrong.
- After you've finished the test, review the answer key to see where you went wrong.

Calculators are not permitted for Grade 3 Common Core Tests

Good Luck!

Common Core GRADE 3 MAHEMATICS REFRENCE MATERIALS

LENGTH

Customary	Metric
1 mile (mi) = 1,760 yards (yd)	1 kilometer (km) = 1,000 meters (m)
1 yard (yd) = 3 feet (ft)	1 meter (m) = 100 centimeters (cm)
1 foot (ft) = 12 inches (in.)	1 centimeter (cm) = 10 millimeters (mm)

VOLUME AND CAPACITY

Customary	Metric
1 gallon (gal) = 4 quarts (qt)	1 liter (L) = 1,000 milliliters (mL)
1 quart (qt) = 2 pints (pt.)	
1 pint (pt.) = 2 cups (c)	
1 cup (c) = 8 fluid ounces (Fl oz)	

WEIGHT AND MASS

Customary	Metric
1 ton (T) = 2,000 pounds (lb.)	1 kilogram (kg) = 1,000 grams (g)
1 pound (lb.) = 16 ounces (oz)	1 gram (g) = 1,000 milligrams (mg)

Time

1 year = 12 months

1 year = 52 weeks

1 week = 7 days

1 day = 24 hours

1 hour = 60 minutes

1 minute = 60 seconds

Common Core Practice Test 1

Mathematics

GRADE 3

Administered *Month Year*

1) What number makes this equation true?

$$13 \times 5 = \square$$

   A. 88

   B. 65

   C. 100

   D. 108

2) Kayla has 120 red cards and 85 white cards. How many more red cards than white cards do Kayla have?

   A. 17

   B. 19

   C. 35

   D. 27

3) A number sentence is shown below.

   $3 \times 5 \square 8 = 120$

   What symbol goes into the box to make the number sentence true?

   A. ×

   B. ÷

   C. +

   D. −

4) Liam had 835 marbles. Then, he gave 432 of the cards to his friend Ethan. After that, Liam lost 116 cards.

Which equation can be used to find the number of cards Eve has now?

A. 835 − 432 + 116 = ____

B. 835 − 432 − 116 = ____

C. 835 + 432 + 116 = ____

D. 835 + 432 − 116 = ____

5) What is the value of "B" in the following equation?

$$43 + B + 7 = 63$$

A. 16

B. 18

C. 22

D. 13

6) There are two different cards on the table.

- There are 3 rows that have 12 red cards in each row.

- There are 21 white cards.

How many cards are there on the table?

A. 25

B. 57

C. 33

D. 99

7) Mason is 15 months now and he usually eats four meals a day. How many meals does he eat in a week?

   A. 36

   B. 40

   C. 28

   D. 48

8) Which of the following list shows only fractions that are equivalent to $\frac{1}{3}$?

   A. $\frac{3}{9}, \frac{5}{15}, \frac{24}{72}$

   B. $\frac{6}{12}, \frac{5}{15}, \frac{9}{27}$

   C. $\frac{3}{9}, \frac{4}{15}, \frac{6}{18}$

   D. $\frac{3}{9}, \frac{5}{10}, \frac{8}{24}$

9) What mixed number is shown by the shaded rectangles?

   A. $2\frac{1}{3}$

   B. $2\frac{3}{4}$

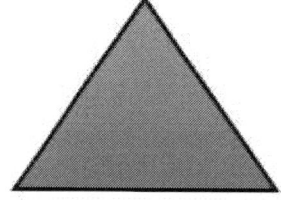

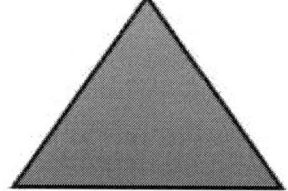

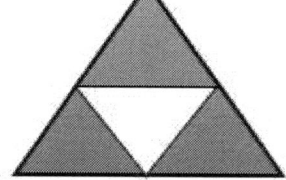

   C. $3\frac{3}{4}$

   D. $2\frac{1}{4}$

10) The perimeter of a square is 32 units. Each side of this square is the same length. What is the length of one side of the square in units?

A. 4

B. 5

C. 6

D. 8

11) Which of the following comparison of fractions is true?

A. $\frac{3}{5} = \frac{9}{15}$

B. $\frac{2}{5} > \frac{4}{10}$

C. $\frac{2}{5} < \frac{4}{10}$

D. $\frac{2}{5} < \frac{2}{10}$

12) The sum of 4 ten thousand, 7 hundred, and 8 tens can be expressed as what number in standard form?

A. 4,780

B. 40,780

C. 40,078

D. 40,708

13) One side of a square is 5 feet. What is the area of the square?

Write your answer in the box below.

14) What is the perimeter of the following triangle?

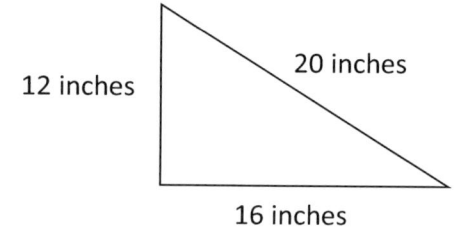

   A. 28 inches

   B. 35 inches

   C. 48 inches

   D. 183 inches

   Triangle with sides labeled 12 inches, 20 inches, and 16 inches.

15) Moe has 460 cards. He wants to put them in boxes of 20 cards. How many boxes does he need?

   A. 20

   B. 21

   C. 22

   D. 23

16) There are 8 rows of chairs in a classroom with 7 chairs in each row. How many chairs are in the classroom?

   A. 45

   B. 56

   C. 54

   D. 63

17) What number goes in the box to make the equation true?

$$\frac{\square}{5} = 2$$

A. 8

B. 10

C. 16

D. 32

18) Which number is represented by A?

13 × A = 169

A. 9

B. 10

C. 13

D. 12

19) What is the perimeter of this rectangle?

A. 12 cm

B. 24 cm

C. 32 cm

D. 64 cm

7 cm

5 cm

20) Nicole has 3 quarters, 5 dimes, and 4 pennies. How much money does Nicole have?

   A. 155 pennies

   B. 125 pennies

   C. 255 pennies

   D. 265 pennies

21) Noah packs 16 boxes with crayons. Each box holds 30 crayons. How many crayons Noah can pack into these boxes?

   A. 480

   B. 540

   C. 680

   D. 720

22) A number sentence such as $88 - x = 28$ can be called an equation. If this equation is true, then which of the following equations is **NOT** true?

   A. $88 - 28 = x$

   B. $88 - x = 28$

   C. $x - 28 = 88$

   D. $x + 28 = 88$

23) There are 6 numbers in the box below. Which of the following list shows only even numbers from the numbers in the box?

$$13, 30, 46, 17, 82, 49$$

A. 13, 30, 46

B. 13, 49, 82

C. 13, 30, 82

D. 30, 46, 82

24) A cafeteria menu had spaghetti with meatballs for $10 and bean soup for $7. How much would it cost to buy five plates of spaghetti with meatballs and two bowls of bean soup?

Write your answer in the box below.

25) Which number correctly completes the number sentence 20 × 45 =?

A. 225

B. 900

C. 1,250

D. 2,250

Common Core Exercise Book – Grade 3

26) There are 60 minutes in an hour. How many minutes are in 5 hours?

   A. 300 minutes

   B. 320 minutes

   C. 360 minutes

   D. 400 minutes

27) Which number correctly completes the number sentence 52 × 14 =?

   A. 550

   B. 660

   C. 728

   D. 990

28) Michael has 845 marbles. What is this number rounded to the nearest ten?

   Write your answer in the box below.

   ☐

29) Use the picture below to answer the question.

   Which fraction shows the shaded part of this square?

   A. $\frac{84}{100}$

   B. $\frac{84}{10}$

   C. $\frac{8.4}{100}$

   D. $\frac{4}{100}$

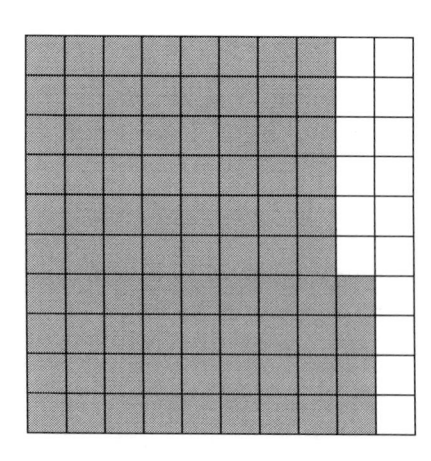

30) Use the table below to answer the question.

Based on their populations, which list of cities is in order from least to greatest?

A. Bryan; Edinburg; Longview; Mission

B. Bryan; Longview; Mission; Edinburg

C. Edinburg; Mission; Longview; Bryan

D. Longview; Edinburg; Mission; Bryan

"This is the end of practice test 1"

Common Core Practice Test 2

Mathematics

GRADE 3

Administered *Month Year*

1) Classroom A contains 7 rows of chairs with 5 chairs per row. If classroom B has three times as many chairs, which number sentence can be used to find the number of chairs in classroom B?

   A. $7 \times 5 + 3$

   B. $7 + 5 \times 3$

   C. $7 \times 5 \times 3$

   D. $7 + 5 + 3$

2) There are 2 days in a weekend. There are 24 hours in day. How many hours are in a weekend?

   A. 48

   B. 96

   C. 168

   D. 200

3) Emily described a number using these clues:

   Three-digit odd numbers that have a 7 in the hundreds place and a 3 in the tens place. Which number could fit Ella's description?

   A. 727

   B. 737

   C. 732

   D. 736

4) A cafeteria menu had spaghetti with meatballs for $8 and bean soup for $7 How much would it cost to buy three plates of spaghetti with meatballs and three bowls of bean soup?

Write your answer in the box below.

☐

5) This clock shows a time after 12:00 PM. What time was it 1 hours and 45 minutes ago?

A. 12:45 PM

B. 1:45 PM

C. 1: 15 PM

D. 12:30 PM

6) A football team is buying new uniforms. Each uniform cost $30. The team wants to buy 12 uniforms.

Which equation represents a way to find the total cost of the uniforms?

A. (30 × 10) + (1 × 12) = 300 + 12

B. (30 × 10) + (10 × 1) = 300 + 10

C. (30 × 10) + (30 × 2) = 300 + 60

D. (12 × 10) + (10 × 20) = 120 + 200

7) Olivia has 93 pastilles. She wants to put them in boxes of 3 pastilles. How many boxes does she need?

   A. 30

   B. 31

   C. 34

   D. 28

8) There are 82 students from Riddle Elementary school at the library on Tuesday. The other 64 students in the school are practicing in the classroom. Which number sentence shows the total number of students in Riddle Elementary school?

   A. 82 + 64

   B. 82 − 64

   C. 82 × 64

   D. 82 ÷ 64

9) Martin earns There are 6 numbers in the box below. Which of the following list shows only odd numbers from the numbers in the box?

13, 30, 24, 18, 73, 39

   A. 13, 24, 18

   B. 13, 39, 73

   C. 13, 30, 24

   D. 24, 18, 30

10) Mia's goal is to save $160 to purchase her favorite bike.

- In January, she saved $46.

- In February, she saved $38.

How much money does Mia need to save in March to be able to purchase her favorite bike?

A. $28

B. $30

C. $52

D. $76

11) Michelle has 84 old books. She plans to send all of them to the library in their area. If she puts the books in boxes which can hold 4 books, which of the following equations can be used to find the number of boxes she will use?

A. $84 + 4 = $ _____

B. $84 \times 4 = $ _____

C. $84 - 4 = $ _____

D. $84 \div 4 = $ _____

12) Which number is made up of 5 hundred, 7 tens, and 6 ones?

A. 5076

B. 576

C. 567

D. 675

13) Elise had 956 cards. Then, she gave 352 of the cards to her friend Alice. After that, Elise lost 250 cards.

Which equation can be used to find the number of cards Elise has now?

A. 956 – 352 + 250 = _____

B. 956 – 352 – 250 = _____

C. 956 + 352 + 250 = _____

D. 956 + 352 – 250 = _____

14) The length of the following rectangle is 9 centimeters and its width is 4 centimeters. What is the area of the rectangle?

A. 12 cm^2

B. 21 cm^2

C. 36 cm^2

D. 22 cm^2

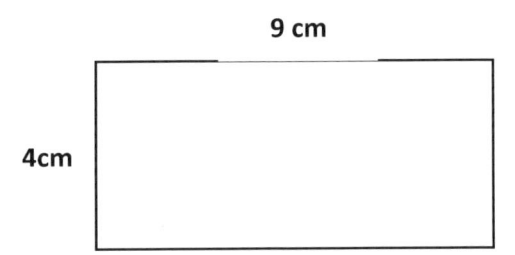

15) Look at the spinner above. On which color is the spinner most likely to land?

A. Red

B. Green

C. Yellow

D. None

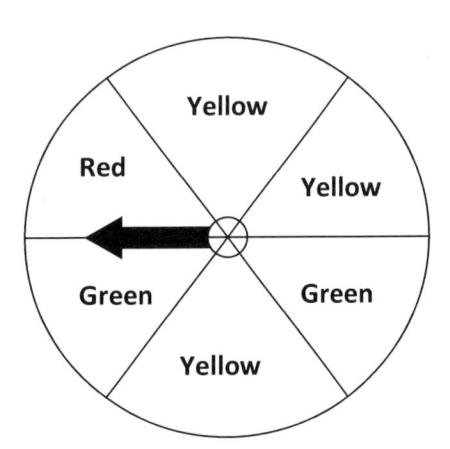

16) A group of third grade students recorded the following distances that they jumped.

23 inches	36 inches	24 inches	28 inches
36 inches	33 inches	25 inches	34 inches
32 inches	28 inches	34 inches	36 inches

What is the distance that was jumped most often?

A. 23

B. 24

C. 32

D. 36

17) Emma flew 3,391 miles from Los Angeles to New York City. What is the number of miles Emma flew rounded to the nearest thousand?

A. 2,000

B. 2,400

C. 2,500

D. 3,000

18) To what number is the arrow pointing?

A. 24

B. 28

C. 30

D. 32

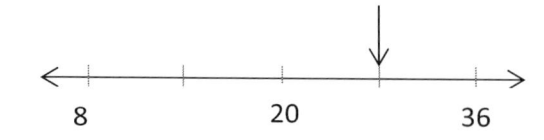

19) A number sentence such as 25 + Z = 92 can be called an equation. If this equation is true, then which of the following equations is not true?

   A. 92 − 25 = Z

   B. 92 − Z = 25

   C. Z − 92 = 25

   D. Z = 67

20) Use the picture below to answer the question. Which fraction shows the shaded part of this square?

   A. $\frac{87}{100}$

   B. $\frac{87}{10}$

   C. $\frac{87}{1,000}$

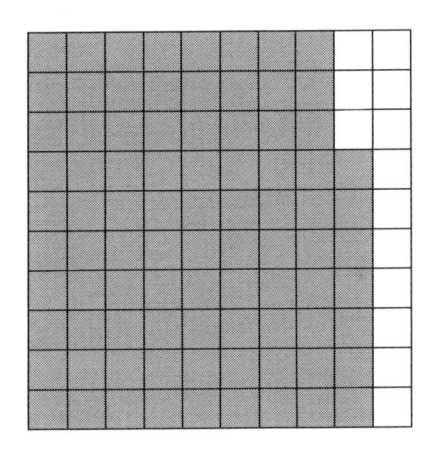

   D. $\frac{8}{100}$

21) Which number correctly completes the number sentence 80 × 35 =?

   A. 350

   B. 900

   C. 1,250

   D. 2,800

22) Which number correctly completes the subtraction sentence

7000 − 858 = _____ ?

A. 6,142

B. 7,452

C. 742

D. 7,458

23) Jason packs 12 boxes with flashcards. Each box holds 30 flashcards. How many flashcards Jason can pack into these boxes?

A. 86

B. 860

C. 530

D. 360

24) Which of the following statements describes the number 26,586?

A. The sum of two thousand, 6 thousand, five hundred, eighty tens, and six ones

B. The sum of sixty thousand, 2 thousand, five hundred, eight tens, and six ones

C. The sum of twenty thousand, 6 thousand, fifty hundred, eighty tens, and six ones

D. The sum of twenty thousand, 6 thousands, five hundreds, eight tens, and six ones

25) The following models are the same size and each divided into equal parts.

The models can be used to write two fractions.

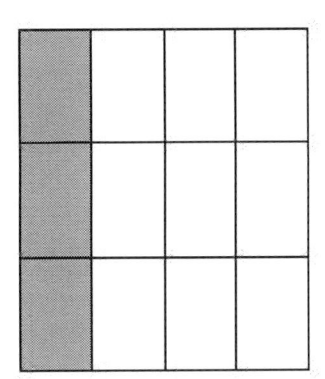

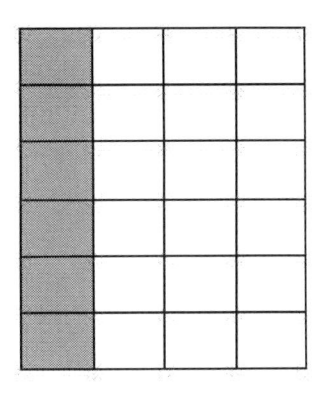

Based on the models, which of the following statements is true?

A. $\frac{3}{12}$ is bigger than $\frac{6}{24}$.

B. $\frac{3}{12}$ is smaller than $\frac{6}{24}$.

C. $\frac{3}{12}$ is equal to $\frac{6}{24}$.

D. We cannot compare these two fractions only by using the models.

26) What is the value of "A" in the following equation?

$$23 + A + 8 = 42$$

A. 10

B. 11

C. 14

D. 20

27) Emily has 144 stickers and she wants to give them to nine of her closest friends. If she gives them all an equal number of stickers, how many stickers will each of Emily's friends receive?

Write your answer in the box below.

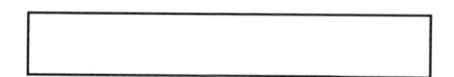

28) Use the models below to answer the question.

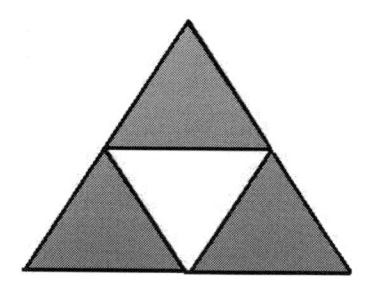

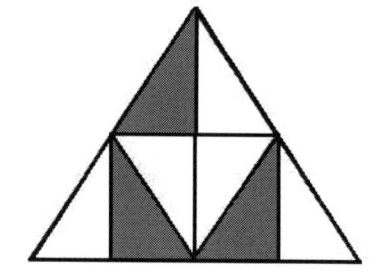

Which statement about the models is true?

A. Each shows the same fraction because they are the same size.

B. Each shows a different fraction because they are different shapes.

C. Each shows the same fraction because they both have 3 sections shaded.

D. Each shows a different fraction because they both have 3 shaded sections but a different number of total sections.

29) Mr. smith usually eats TWO meals a day. How many meals does he eat in a week?

   A. 21

   B. 14

   C. 28

   D. 30

30) What is the value of A in the equation $56 \div A = 8$?

   A. 2

   B. 6

   C. 7

   D. 9

"This is the end of the practice test 2"

Answers and Explanations

Common Core Practice Tests

Answer Key

✳ Now, it's time to review your results to see where you went wrong and what areas you need to improve!

Common Core - Mathematics							
Practice Test - 1				**Practice Test - 2**			
1	B	11	A	21	D	1	C	11	D	21	D
2	C	12	B	22	C	2	A	12	B	22	A
3	A	13	25	23	D	3	B	13	B	23	D
4	B	14	C	24	64	4	45	14	C	24	D
5	D	15	D	25	B	5	D	15	C	25	C
6	B	16	B	26	A	6	C	16	D	26	B
7	C	17	B	27	C	7	B	17	D	27	16
8	A	18	C	28	850	8	A	18	B	28	D
9	B	19	B	29	A	9	B	19	C	29	B
10	D	20	B	30	D	10	D	20	A	30	C

Practice Test 1

Common Core - Mathematics

Answers and Explanations

1) Answer: B.

13 × 5 = 65

2) Answer: C.

To find the answer subtract 85 from 120. The answer is (120 – 85) = 35.

3) Answer: A.

3 × 5 = 15. Then:

3 × 5 ☐ 8 = 120

15 ☐ 8 = 120 ⇒ 120 = 15 × 8

4) Answer: B.

Liam gave 432 of his marbles to his friend. Now he has 835 – 432 = 403

He lost 116 of his marbles. Now, he has 403 – 116 = 287 or (835 – 432 – 116).

5) Answer: D.

43 + B + 7 = 63 ⇒ 50 + B = 63 ⇒ B = 63 – 50 = 13

6) Answer: B.

3 rows that have 12 red cards in each row contain: 3 × 12 = 36 red cards

And there are 21 white cards on table. Therefore, there are 36 + 21 = 57 cards on table.

7) Answer: C.

If Mason eats 4 meals in 1 day, then, in a week (7days) he eats (7 × 4 = 28) meals.

8) Answer: A.

All these fractions; $\frac{3}{9}, \frac{5}{15}, \frac{24}{72}$ are equivalent to $\frac{1}{3}$.

9) Answer: B.

This shape shows 2 complete shaded rectangle and 3 parts of a triangle divided into 4 equal parts. It is equal to $2\frac{3}{4}$.

10) Answer: D.

Perimeter of the square is 32. Then:

32 = 4 × side ⇒ side = 8

Each side of the square is 8 units.

11) Answer: A.

Simplify $\frac{9}{15}$ that's equal to $\frac{3}{5}$. Only option A is correct.

12) Answer: B.

4 ten thousand = 40,000

7 hundred = 700

8 tens = 80

Add all: 40,000 + 700 + 80 = 40,780

13) Answer: 25.

To find the area of a square, multiply one side by itself.

Area of a square = (side) × (side) = 5 × 5 = 25

14) Answer: C.

To find the perimeter of the triangle, add all three sides.

Perimeter = 12 + 16 + 20 = 48 inches

15) Answer: D.

Moe wants to put 460 cards into boxes of 20 cards. Therefore, he needs (460 ÷ 20 =) 23 boxes.

16) Answer: B.

8 rows of chairs with 7 chairs in each row means: 8× 7 = 56 chairs in total.

17) Answer: B.

We need to find a number that when divided by 5, the answer is 2. Therefore, we are looking for 10.

18) Answer: C.

A = 169 ÷ 13 ⇒ A = 13

Common Core Exercise Book – Grade 3

19) Answer: B.

Use perimeter of rectangle formula.

Perimeter = 2 × length + 2 × width ⇒ P= 2 × 5+ 2 × 7 = 10 +14 = 24 cm

20) Answer: B.

3 quarters = 3 × 25 pennies = 75 pennies

5 dimes = 5 × 10 pennies = 50 pennies

In total Nicole has 125 pennies

21) Answer: D.

16 × 30 = 480

22) Answer: C.

$88 - x = 28$

Then, $x = 88 - 28 = 60$

Let's review the equations provided:

A. $88 - 28 = x$ This is true!

B. $88 - x = 28$ This is true!

C. $x - 28 = 88$ This is NOT true!

D. $x + 28 = 88$ This is true!

23) Answer: D.

Even numbers always end with a digit 0, 2, 4, 6 or 8.

Therefore, numbers 30, 46, 82 are the only even numbers.

24) Answer: 64.

5 spaghetti with meatballs cost: 5 × $10 = $50

2 bowls of bean soup cost: 2 × $7 = $14

5 spaghetti with meatballs + 2 bowls of bean soup cost: $50 + $ 14 = $64

25) Answer: B.

20 × 45 = 900

26) Answer: A.

1 hour = 60 minutes

5 hours = 5 × 60 minutes ⇒ 5hours = 300 minutes

WWW.MathNotion.Com

27) Answer: C.

52 × 14 = 728

28) Answer: 850.

We round the number up to the nearest ten if the last digit in the number is 5, 6, 7, 8, or 9.

We round the number down to the nearest ten if the last digit in the number is 1, 2, 3, or 4.

If the last digit is 0, then we do not have to do any rounding, because it is already rounded to the ten.

Therefore, rounded number of 845 to the nearest ten is 850.

29) Answer: A.

The table is divided into 100 equal parts. 88 parts of these 100 parts are shaded. It means $\frac{84}{100}$.

30) Answer: D.

Longview city with 83,287 has the least population. Bryan, Mission and Edinburg are other cities in order from least to greatest.

Practice Test 2
Common Core - Mathematics
Answers and Explanations

1) Answer: C.

Classroom A contains 7 rows of chairs with four chairs per row. Therefore, there are (7 × 5 =) 35 chairs in Classroom A. Classroom B has three times as many chairs. Then, there are (7 × 5 × 3) chairs in Classroom B.

2) Answer: A.

1 day: 24 hours

2 days = 2× 24 = 48 hours

3) Answer: B.

Three-digit odd numbers that have a 7 in the hundreds place and a 3 in the tens place is 737. 732 and 736 are even numbers.

4) Answer: 45.

3 plates of spaghetti with meatballs cost: 3 × $8 = $24

3 bowls of bean soup cost: 3 × $7 = $21

3 plates of spaghetti with meatballs + 3 bowls of bean soup cost: $24 + $21 = $45

5) Answer: D.

The clock shows 2:15 PM. One hour before that was 1:15 PM. 45 minutes before that was 12:30 PM.

6) Answer: C.

The Football team buys 12 uniforms that each uniform cost $30. Therefore, they should pay (12 × $30 =) $360.

Choice C is the correct answer.

(30 × 10) + (30 × 2) = 300 + 60 = 360

7) Answer: B.

Olivia wants to put 93 pastilles into boxes of 3 pastilles.

Common Core Exercise Book – Grade 3

Therefore, she needs (93 ÷ 3 =) 31 boxes.

8) Answer: A.

To find the total number of students in Riddle Elementary School, add 82 with 64.

9) Answer: B.

An easy way to tell whether a large number is odd or even is to look at its final digit. If the number ends with an odd digit (1, 3, 5, 7, or 9), then it's odd. On the other hand, if the number ends with an even digit or 0 (0, 2, 4, 6 or 8), then it is even.

13, 39 and 73 ends with odd digit, therefore, they are odd numbers.

10) Answer: D.

Mia saved $46 and $38. Therefore, she has $84 now.

$160 - $84 = $76. She needs to save 76.

11) Answer: D.

Michelle puts 84 books in 4 boxes. Therefore, 84 ÷ 4 formula is correct.

12) Answer: B.

To find the number, put 5 for hundreds place, 7 for tens place, and 6 for one's place.

Then, you will get: 500 + 70 + 6 = 576

13) Answer: B.

Elise gave 352 of her 956 cards to her friend. Therefore, she has 956 – 352 cards now. Then she lost 250 cards. Now, she has (956 – 352 – 250) = 354 cards

14) Answer: C.

Use area formula of a rectangle:

Area = length × width

Area = 4cm × 9cm = 36 cm^2

15) Answer: C.

The chance of landing on yellow is 3 out of 6.

The chance of landing on red is 1 out of 6.

The chance of landing on green is 2 out of 6.

The chance of landing on yellow red is more than the chance of landing on other colors.

WWW.MathNotion.Com

16) Answer: D.

36 is the most frequent number in the table.

17) Answer: D.

The number 3,391 rounded to the nearest thousand is 3,000.

18) Answer: B.

The arrow shows a number between two numbers 20 and 36. $(36 - 20 = 16, 16 \div 2 = 8) \Rightarrow 20 + 8 = 28$

Therefore, the answer is 28.

19) Answer: C.

25 + Z = 92. Then, Z = (92 – 25=) 67.

All these equations are true:

92 – 25 = Z

92 – Z = 25

Z = 67

But this equation is not true: Z – 92 = 25

20) Answer: A.

The table is divided into 100 equal parts. 87 parts of these 100 parts are shaded. The shaded part is equal to $\frac{87}{100}$.

21) Answer: D.

$80 \times 35 = 2,800$

22) Answer: A.

$7,000 - 858 = 6,142$

23) Answer: D.

To find the answer, multiply 12 by 30.

$12 \times 30 = 360$

24) Answer: D.

26,586 is the sum of:

$20,000 + 6,000 + 500 + 80 + 6$

25) Answer: C.

The first model is divided into 12 equal parts. 3 out of 12 parts are shaded. That means $\frac{3}{12}$ which is equal to: $\frac{1}{4}$

The second model is divided into 24 equal parts. 6 out of 24 parts are shaded. That means $\frac{6}{24}$ which is equal to: $\frac{1}{4}$

26) Answer: B.

A = 42 − 23− 8 = 11

27) Answer 16.

144 ÷ 9 = 16

28) Answer: D.

The first model from left is divided into 4 equal parts. 3 out of 4 parts are shaded. The fraction for this model is $\frac{3}{4}$. The second model is divided into 8 equal parts. 3 out of 8 parts are shaded. Therefore, the fraction of the shaded parts for this model is $\frac{3}{8}$. These two models represent different fractions.

29) Answer: B.

2 meals a day, means (2 × 7 =) 14 meals a week.

30) Answer: C.

A = 56 ÷ 8 = 7

"End"

Made in the USA
Columbia, SC
09 December 2022